a survey of LIVING PRIMATES and Their ANATOMY

Friderun Ankel-Simons
Duke University

a survey of LIVING PRIMATES and Their ANATOMY

Macmillan Publishing Co., Inc.
New York

Collier Macmillan Publishers
London

Copyright © 1983 Friderun Ankel-Simons

Printed in the United States of America

All rights reserved. No part of this book may be reproduced or transmitted in any form or by any means, electronic or mechanical, including photocopying, recording, or any information storage and retrieval system, without permission in writing from the Publisher.

Macmillan Publishing Co., Inc.
866 Third Avenue, New York, New York 10022
Collier Macmillan Canada, Inc.

Library of Congress Cataloging in Publication Data

Ankel-Simons, Friderun.
 A survey of living primates and their anatomy.

 Bibliography: p.
 Includes index.
 1. Primates—Anatomy. I. Title.
QL737.P9A594   1983   599.8044   82-4677
ISBN 0-02-303500-5

Printing: 1 2 3 4 5 6 7 8   Year: 3 4 5 6 7 8 9 0

ISBN 0-02-303500-5

*This book is dedicated to
my dear friend Elsa Dubois,
my husband Elwyn,
our children, Cornelia and Verne,
and to the memory of
my brother, Cornelius Ankel*

Preface

Many people give various kinds of help to an author during the preparation of a book. Such aid—both explicit and implicit—deserves the author's thanks.

First of all, I would like to thank my students, whose inquisitive questions have always been an invaluable stimulus that has taught me more than have many teachers.

Furthermore, my gratitude goes to all the colleagues who assisted me in various ways and especially to those who read and contributed their criticism to parts of the book, namely Fredericka Oakley, Matt Cartmill, Bert Covert, Dieter Glaser, Andy Hamilton, David Pilbeam, Montrose Moses, Patricia Poorman, Len Radinksy, and Ian Tattersall. Rich Kay also helped with some library problems. My husband Elwyn Simons, edited the manuscript and added much invaluable knowledge and advice. Ruth Nix was helpful with editorial matters.

I am also very grateful to all those who contributed photographs: Alison Richard, Ken Glander, Dieter Glaser, Christian Schmidt, Michael Stuart, and Heinrich Sprankel.

Lastly, my dear friend Elsa Dubois gave me support in so many ways that a very special word of thanks goes to her.

One final word about the scope of this book seems appropriate here: Because both the primate fossil record and the details of the behavior of living primates have already been covered by Elwyn Simons and Alison Jolly in their respective contributions to the Macmillan Series in Physical Anthropology, no attempt is made here to duplicate any of the information included in their two volumes.

F.A.S.

Contents

Taxonomic List of Extant Primates — xi

1. /History and Objectives of Primatology — 1

2. /Definition of Order Primates — 9

3. /Teeth — 77

4. /Skull — 113

5. /Brain — 136

6. /Postcranial Skeleton — 151

7. /Sense Organs and Viscera   200

8. /Placentation, Reproduction, and Growth   228

9. /Locomotion   244

10. /Chromosomes and Blood Groups   258

Appendix   283

Bibliography   286

Index   303

Taxonomic List of Extant Primates

Intercedents Insectivora—Primates

Infraorder Tupaiiformes

Family Tupaiidae

Subfamily Ptilocercinae
    Ptilocercus   *Ptilocercus lowii*
    (pen tailed tree shrew)

Subfamily Tupiinae
    Tupaia   *Tupaia glis*
        (tree shrews)   *Tupaia picta*
                               Tupaia carimatae
                               Tupaia nicobarica
                               Tupaia javanica
                               Tupaia minor
                               Tupaia gracilis
    Tana   *Tupaia (Tana) tana*
                               Tupaia (tana) dorsalis
                               Tupaia (Anathana) ellioti
    Anathana   *Dendrogale murina*
    Dendrogale   *Dendrogale melanura*
    Urogale   *Urogale everetti*

Order Primates

Suborder Prosimii

Infraorder Lemuriformes

Superfamily Lemuroidea

Family Lemuridae

Subfamily Lumurinae
    Lemur   *Lemur catta*
        (lemurs)   *Lemur fulvus*
                                     Lemur macaco
                                     Lemur mongoz
                                     Lemur rubriventer
    Varecia   *Lemur* or *Varecia variegatus*
      (variegated lemur)
    Hapalemur   *Hapalemur griseus*
      (gentle lemur)   *Hapalemur simus*
    Lepilemur   *Lepilemur mustelinus*
      (sportive lemur)

Subfamily Cheirogaleinae
    Cheirogaleus   *Cheirogaleus major*
      (dwarf lemurs)   *Cheirogaleus medius*
    Allocebus   (*Cheirogaleus*) *Allocebus trichotis*
  (hairy eared dwarf lemur)
    Microcebus   *Microcebus murinus*
    Mirza   *Mirza coquereli*
      (mouse lemurs)
    Phaner   *Phaner furcifer*
      (forked lemur)

Family Indriidae
Subfamily Indriinae
    Propithecus   *Propithecus diadema*
      (sifakas)   *Propithecus verreauxi*
    Indri   *Indri indri*
      (indri)
    Avahi   *Avahi laniger*
(woolly lemur)

Family Daubentoniidae
    Daubentonia   *Daubentonia madagascariensis*
      (aye aye)

Infraorder Lorisiformes
Superfamily Lorisoidea

Family Lorisidae
    Loris   *Loris tardigradus*
        (slender loris)
    Nycticebus   *Nycticebus coucang*
        (slow loris)   *Nycticebus pygmaeus*
                                          Nycticebus intermedius
    Arctocebus   *Arctocebus calabarensis*
        (golden potto)
    Perodicticus   *Perodicticus potto*
        (potto)
Subfamily Galaginae
    Galago   *Galago senegalensis*
        (bush babies)   *Galago crassicaudatus*
                                          Galago alleni

Subgenus
    Euoticus   *Galago (Euoticus) elegantulus*
    (needle-clawed bush baby) *Galago (Euoticus) inustus*
    Galagoides   *Galago (Galagoides) demidovii*
        (dwarf bush baby)

Infraorder Tarsiiformes

Superfamily Tarsioidea

Family Tarsiidae

Subfamily Tarsiinae
    Tarsius   *Tarsius syrichta*
        (tarsiers)   *Tarsius bancanus*
                                       Tarsius spectrum

Suborder Anthropoidea

Infraorder Platyrrhini

Superfamily Ceboidea

Family Cebidae

Subfamily Aotinae
    Aotus   *Aotus trivirgatus*
        (night monkey)

Callicebus
(titi monkeys)

Callicebus torquatus
Callicebus moloch
Callicebus personatus

Subfamily Pithecinae
Cacajao
(uakaris)

Cacajao melanocephalus
Cacajao rubicundus
Cacajao calvus

Pithecia
(sakis)

Pithecia pithecia
Pithecia monacha

Chiropotes
(bearded sakis)

Chiropotes satanas
Chiropotes albinasus

Subfamily Alouattinae
Alouatta
(howler monkeys)

Alouatta villosa
Alouatta seniculus
Alouatta fusca
Alouatta caraya
Alouatta belzebuth

Subfamily Cebinae
Cebus
(capuchin monkeys)

Cebus apella
Cebus capucinus
Cebus albifrons
Cebus nigrivittatus

Saimiri
(squirrel monkeys)

Saimiri sciureus
Saimir oerstedii

Subfamily Atelinae
Ateles
(spider monkeys)

Ateles paniscus
Ateles belzebuth
Ateles fusciceps
Ateles geoffroyi

Brachyteles
(woolly spider monkey)

Brachyteles arachnoides

Lagothrix
(woolly monkeys)

Lagothrix lagothricha
Lagothrix flavicauda

Family Callitrichidae

Subfamily Callimiconinae
Callimico
(Goeldi's marmoset)

Callimico goeldii

Callithrix
(marmosets)

Callithrix argentata
Callithrix aurita

Taxonomic List of Extant Primates

                                  Callithrix flaviceps
                                  Callithrix geoffroyi
                                  Callithrix jacchus
                                  Callithrix penicillata
                                  Callithrix humeralifer
                                  Callithrix chrysoleuca

    Cebuella   *Cebuella pygmaea*
        (pygmy marmoset)

    Saguinus   *Saguinus tamarin*
        (tamarins)   *Saguinus devillei*
                                    Saguinus fuscicollis
                                  Saguinus fuscus
                                  Saguinus bicolor
                                  Saguinus martinsi
                                  Saguinus leucopus
                                  Saguinus inustus
                                  Saguinus midas
                                  Saguinus imperator
                                  Saguinus mystax

    Leontideus   *Leontideus rosalia*
        (lion marmosets) *Leontideus chrysomelas*
                                  Leontideus chrysopygus

    Oedipodmidas   *Oedipomidas oedipus*
        (pinche marmosets) *Oedipomidas geoffroyi*

Infraorder Catarrhini

Superfamily Cercopithecoidea

Subfamily Cercopithecinae

    Macaca   *Macaca sylvana*
        (macaques)   *Macaca mulatta*
                                  Macaca fascicularis
                                  Macaca maurus
                                  Macaca fuscata
                                  Macaca speciosa
                                  Macaca sinica
                                  Macaca radiata
                                  Macaca cyclopsis
                                  Macaca silena
                                  Macaca nemestrina
                                  Macaca assamensis
                                  Macaca arctoides

    Cynopithecus   *Cynopithecus (Macaca) niger*
        (Celebes black ape)

Cercocebus (mangabeys)	*Cercocebus albigena* *Cercocebus aterrimus* *Cercocebus torquatus* *Cercocebus atys* *Cercocebus galeritus*
Papio (baboons)	*Papio hamadryas* *Papio anubis* *Papio cynocephalus* *Papio papio* *Papio ursinus*
Papio (Mandrillus) (mandrills)	*Papio (Mandrillus) sphinx* *Papio (Mandrillus) leucophaeus*
Theropithecus (gelada baboon)	*Theropithecus gelada*
Cercopithecus (guenons, vervets)	*Cercopithecus aethiops* *Cercopithecus pygerythrus* *Cercopithecus sabaeus* *Cercopithecus cephus* *Cercopithecus diana* *Cercopithecus lhoesti* *Cercopithecus preussi* *Cercopithecus hamlyni* *Cercopithecus petaurista* *Cercopithecus mitis* *Cercopithecus albogularis* *Cercopithecus mona* *Cercopithecus campbelli* *Cercopithecus wolfi* *Cercopithecus denti* *Cercopithecus pogonias* *Cercopithecus neglectus* *Cercopithecus nictitans* *Cercopithecus ascanius* *Cercopithecus erythrotis* *Cercopithecus erythrogaster*
Cercopithecus (Allenopithecus) (Allen's swamp monkey)	*Cercopithecus (Allenopithecus) nigroviridis*
Cercopithecus (Miopithecus) (talapoin)	*Cercopithecus (Miopithecus) talapoin*
Erythocebus (patas)	*Erythrocebus patas*

Taxonomic List of Extant Primates

Subfamily Colobinae
    Presbytis
        (langurs)

Presbytis entellus
Presbytis senex
Presbytis johnii
Presbytis aygula
Presbytis melalophos
Presbytis frontatus
Presbytis rubicundus
Presbytis cristatus
Presbytis obscurus
Presbytis phayrei
Presbytis francoisi
Presbytis potenziani
Presbytis pileatus
Presbytis geei

    Pygathrix
        (douc langur)

Pygathris nemaeus

    Rhinoptithecus
        (snubnosed langurs)

Rhinopithecus roxellanae
Rhinopithecus avunculus

    Simias
        (Pagai island langur)

Simias concolor

    Nasalis
        (proboscis monkey)

Nasalis larvatus

    Colobus
        (guerezas)

Colobus polykomos
Colobus guereza
Colobus (Procolobus) verus
Colobus (Piliocolobus) badius
Colobus (Piliocolobus) kirkii

Superfamily
Family Hylobatidae
Subfamily Hylobatinae
    Hylobates
        (gibbons)

Hylobates lar
Hylobates agilis
Hylobates moloch
Hylobates hoolok
Hylobates concolor
Hylobates klossii

    Symphalangus
        (siamang)

Symphalangus syndactylus

Family Pongidae
Subfamily Ponginae
    Pongo   *Pongo pygmaeus*
        (orangutan)
    Pan   *Pan troglodytes*
        (chimpanzees)   *Pan paniscus*
    Gorilla   *Gorilla gorilla*
        (gorilla)

Superfamily Hominoidea
Family Hominidae
    Homo   *Homo sapiens*
        (human)

a survey of
LIVING PRIMATES
and Their
ANATOMY

chapter 1
History and Objectives of Primatology

The science of primatology is concerned with the study of those mammals most closely related to human beings. These mammals constitute the Order Primates, an order that includes four main living groups: prosimians, monkeys, apes, and man. Some more distantly related forms such as tree shrews, colugos, and elephant shrews may belong in separate orders but are also of interest. Today, primatology is an important subdiscipline of biology. Even though diverse studies on primates are a relatively recent scientific development, the literature in this field has been expanding rapidly since the late 1950's.

Thomas Henry Huxley was perhaps the first scientist to apply, in detail, Darwin's theory of evolution by natural selection to the interpretation of the comparative biology of humans and apes. In his 1863 essay entitled "Man's Place in Nature," Huxley first dealt with many of the topics that have remained important in primatology up to the present day. Such topics include the position of human beings among the primates and the nature of our descent from increasingly "primitive" grades of organization.

For those who engage in research on the primates, primatology has never seemed more relevant than at present. Arising from diverse beginnings, its subdisciplines are becoming more closely integrated. As numerous more precise data about primates are published, the subject is making increasing contributions to biological studies as a whole.

Within the biological sciences, primatology is closest to physical anthropology, a discipline that is specifically concerned with analysis of our own species: *Homo sapiens*—the only species capable of self-understanding.

In spite of centuries of developing human self-interest, many aspects of human biology and primatology have only recently been explored. As a discipline concerned mainly with one species, *Homo sapiens*, anthropology has a unique coincidence of subject and object, but most physical anthropologists also study non-human primates as analogues to ourselves. Humans are fascinated by their near relatives; we continue to be amused, even shocked, by the many parallels to ourselves.

One could compile a lengthy account of references to primates in literature, but here a brief outline must do. In the fourth century B.C., the philosopher Aristotle, in *Historia Animalium*, initially divided monkeys into three main groups. These were: 1. the *pithekoi*, forms with reduced tails, 2. the *keboi*, forms with unreduced tails, and 3. the *kynokephaloi*, dog-headed forms, i.e., baboons. In his *Natural History*, Pliny the Elder observed that the primates are much like man. Later, Galen of Pergamon dissected monkeys and apes and pointed out that they closely resembled man in the bony skeleton and in the intestinal, muscular, nervous, and vascular systems. He wisely advised his students to study the primates in order to gain a better understanding of the anatomy of man.

Marco Polo, who traveled widely in the Orient in the thirteenth century, described strange, small, manlike creatures. This was perhaps the earliest reference to lesser apes or gibbons. From Marco Polo's time on, scholars in Europe showed an increasing interest in the natural world. By the sixteenth century, Conrad Gessner in Switzerland reviewed all he could find about primates for his *Natural History*. This outstanding early work reflects, together with a certain credulousness and the superstition characteristic of those times, the inception of ecstatic feelings about the wonders of the natural world. In 1699 an English scholar, Edward Tyson, published the first study of the anatomy of an ape, basing his work on the body of a "pygmie" from Angola that was later understood to be that of a young chimpanzee. In spite of its early date, this study was remarkably accurate. In the two hundred years after the study, many descriptions of monkeys and apes were published in Europe. Their authors included the well-known natural historians Blumenbach, Buffon, Cuvier, Erxleben, Illiger, Owen, Pennant, and Geoffroy Saint-Hillaire, all of whom added significantly to knowledge of primates.

Attempts to organize the taxonomy of primates began in Sweden in 1758 when the naturalist Linnaeus published a remarkable work. This was the tenth edition of his famous book, *Systema Naturae*, in which he named one order of mammals Primates. In this order he placed, together with man, a genus of ape, of monkey, of lemur, and of bat. Twenty-three years before this publication, in the first edition of his work, he had already grouped humans, apes, and monkeys together and also (with unintentional humor) the sloths. These he had ranked together in one group, the "Anthropomorpha" or manlike creatures. For his objective, Linnaeus systematically ranked animals only according to their similarities and drew no conclusions about man's place in nature. Nevertheless, his bold step in uniting man with animals caused protest, and others soon began to reassert the uniqueness of humans by separating them as distantly as possible from all other living organisms.

Thus, Johann Friedrich Blumenbach, in 1791, separated men from an embarrassingly close relationship to apes by creating two different orders. One was the order "Biamana" (two-handed) for man, and a second, the order "Quadrumana" (four-handed) for the remaining primates. The same distinction was made by Baron Cuvier nine years later, and the use of both terms persisted for nearly 100 years thereafter. Differing with this usage, Illiger (1811) took as the central concept of his systematics the uprightness of humans and established for them the order "Erecta." Owen (1868) believed that the difference between us and other primates was great enough to make a much higher category in the animal kingdom for our kind. He coined for us the subclass "Archencephala," those with the most advanced kind of brains.

Beginning in 1859, Darwin brought a fresh point of view to the discussion of our relationship to other animals. For him the similarities between different organisms were due neither to design nor to chance. He recognized that the relationships of living things to each other showed that the similarities between animals are due to common descent. Darwin thereby made a critical push toward a new kind of biological thinking, although he avoided, at that time, the implications of natural selection as the basis for the origin of *Homo sapiens*.

A few years later, as we have already mentioned, Thomas Henry Huxley (1863) made the significant step of considering humans in the context of animal evolution in his work, *Man's Place in Nature*, in which the exact relationships of humans and apes were stressed. Finally, Charles Darwin (1871), in *The Descent of Man*, made an elaborate study comparing human and animal. From then on, many scientists throughout the last decades of the nineteenth century and during the early part of this century dealt with the ties between humans and the other primates as the full impact of the biological nature of humans developed. Together, these publications have shown that primatology contributes a necessary background for understanding the main stages of human evolution.

Because of the identity of subject and object, a high level of subjectivity characterizes much that has been done in anthropology and that has come into the study of our species. Primatology as a whole provides new and better sources of objective information that should help to clarify some of the phases of understanding human evolution that have been controversial in the past. Considered as merely one species of the Linnaean Order Primates, *Homo sapiens* can be dealt with more objectively.

Biology is essentially a comparative science. The relationships of organisms one to another, their similarities and their distinctions, are the bases of contrast. Of itself, a single biological object has no context. Because it is impossible to avoid recognizing the many similarities between humans and apes, the study of humans, considered as animals, gains both strength and objectivity in a comparative approach. Were it not for the uniqueness of humans in the natural world, there would be less importance to primatology; there would be no more interest in this particular mammalian order than in the others. As many of the strengths of present-day physical anthropology are derived from primatology, primatology in turn is dependent on understanding other animals, especially other nonprimate mammals. Thus, primatology cannot be taken as an entirely self-contained field.

In considering more recent advances in primatology, one thinks automatically of such leading figures in the field as Adolph Hans Schultz of Zurich or of the English anatomist Sir Wilfrid Le Gros Clark, both of whom, from early in this century, began publishing a series of fundamental contributions to primatology. Even so, the term "primatology" seems to have been first used in print as recently as 1941 by T.C. Ruch. Anthropologists and paleontologists, making use of the analogies to be drawn from the study of recent primates, have endeavored to reconstruct the natural history of humans and primates.

This history often has had to be interpreted from meager evidence. Primate paleontological, molecular, and behavioral research provides the most important kinds of clues to this particular type of study. The late development of this field is indicated by the fact that one of the first university courses in primatology was taught by E.L. Simons as recently as 1959.

Form and function are closely interrelated in bones as well as in teeth. By studying the movements of living primates, we can begin to identify the relationships between form and function in them. With caution, and within definite limits, these functional interpretations can (by analogy) be applied to fossil forms in order to reconstruct the function of extinct animals. This can only be done effectively when we know as much as possible about the life style of present-day species. In fact, it would be an exaggeration to imply that we know well the biology of most living primates, but nonetheless the groundwork has been laid. Fundamental studies on the locomotor behavior of living primates are now being undertaken in increasing numbers. The discussion of terminology for 'basic locomotor types continues unabated. For example, it long seemed impossible to reconstruct the locomotor behavior of earliest hominids. Now, with new comparative knowledge gained from other living primates, we can approach the problem. There has been a lengthy debate about whether the ancestors of man, before they became upright walkers, were brachiators living in the forest canopy or alternatively whether they were quadrupedal branch runners and climbers. It is clear from fossil finds and comparative studies made in recent years that we have gained clarification about these alternatives. The supposition that wild gorillas and chimpanzees were brachiators (resembling the small Asiatic gibbons) long survived without foundation from field observation. More recently, field observations have shown that chimpanzees as adults rarely swing by their arms and that gorillas virtually never do this.

Comparative researches on a broad range of primates have shown that we are not only very different from other primates in bodily proportions and in the construction of skull and face but in our complex way of life. However, biochemical findings of the last few years indicate great similarities between humans and the African apes. The intrinsic complexity of a single individual increases as the structural and behavioral organization of animals becomes more advanced. This is especially true for monkeys and apes. It was believed for a long time that individual monkeys within a species were nearly the same if not identical to each other. Thus, it was not considered wrong to generalize from findings based on one or two individual monkeys to the whole species. We now see that the high degree of present human structural and behavioral variability extends not only to other primate species but to nonprimates as well.

Thus, it is clear that in the high individual variability of primates, we have a rich source of possible error in the interpretation of fossils. By understanding the range of variability to be found within and between species of related living primates, we can avoid this error. This knowledge of variability has become a principal basis for the latest taxonomic revisions of fossil finds. In general, such revisions suggest grouping of fossils that previously had separate names, and this in turn makes the picture of the evolutionary history of primates easier to grasp.

The course of human evolution is now documented by a large number of fossils, and there will surely be collections of many more such fossils if the present search for human forerunners is continued. The rough outline of the successive phases in the history of humans during the last two million years can now be drawn with general agreement. Recent finds have probably doubled this age to four million for the earliest *Australopithecus*, but the period from two to four million years ago is not so clearly understood. *Australopithecus* has a skull that is outwardly more reminiscent of the apes than of modern man, but the teeth are not ape-like. There is definite evidence that by about three and a half million years ago at least, some hominids had already achieved an upright gait.

Darwin and several of his contemporaries already recognized that humans and apes were near relatives. Recent research has shown that this resemblance is closely built-in. Today, this knowledge is reinforced by the most modern methods of cytology, serology, and molecular biology. It is clear that many characteristics of the human organism differ only quantitatively from other primates. It is, therefore, more the exaggeration of certain characteristics in man rather than qualitative differences that makes him distinct.

An hypothesis of the Dutch scientist Bolk, who in 1926 suggested that man is nothing but an ape that has retained infant proportions into adult stages, has received broad circulation. This speculation is fascinating only at first glance. Examples of slowing development in the ontogeny of humans, which would substantiate Bolk's theory, can easily be found: For example, the late fusion of the sutures between the bones of the brain case. Equally, it is easy to find examples of speeding up rather than slowing down in human embryonic development, such as the early fusion of the elements of the sternum. When one has the advantage of knowing the developmental history of a broad spectrum of primates, it is clear that the developmental difference between humans and the other primates is achieved by a combination of acceleration and retardation. New research on comparative behavior of primates, both in adult societies and during the individual's behavioral growth, has changed our thinking. The dependency on the mother up to puberty was thought to be unique to man. Today we know that here too we have only differences of degree between man, apes, and monkeys. Newborn apes show as much need (but for a shorter period) for the mother's care. Although juvenile development, the learning period, the onset of puberty, and the following phases of life are absolutely shorter in apes than in modern man, all follow the same fundamental sequence. Even newborn monkeys, which are more self-sufficient than infant humans and apes, cannot survive in the wild without the mother and without the whole troop in many cases. Because young monkeys have much to learn and because, within the troop, experiences are passed from generation to generation by example, we see here the beginning of different traditions varying from troop to troop within the same species.

Young monkeys have to practice activities that will be important for their integration into adult social life. During play they come to understand their physical abilities. They learn how to defend themselves, how to help themselves in difficult situations, and how to escape. Juveniles isolated from their mothers and from the group do not develop the proper behavioral repertoire for social integration. Such monkeys cannot later develop the capacity for complex social interactions with conspecifics.

Thus, the mother and her care are very important during the first stages of life. After this, play with other infants is necessary for later behavioral development. Such findings are important in understanding human behavior and in all attempts to reconstruct possible early human or prehuman behavior. Nevertheless, it is simplistic to suppose that behavioral observations on the nonhuman primates can be used directly to interpret early man. Play in monkeys, in human children, or among human ancestors may have, or have had, somewhat different functions.

Thus, comparative research among the primates can demonstrate that human ontogeny after birth indeed has a certain uniqueness. All phases of life—childhood, youth, adulthood, and old age—are absolutely longer in humans than in any other present-day primate. The difference of *Homo sapiens* here is particularly marked in the later maturation of individuals together with the continued accumulation of individual wisdom and knowledge, a process upon which much of man's civilization depends. This long period of old age is a new development in organic evolution. We all know that the human life span long outlasts the reproductive period, and in fact much of what it means to be human depends on this ability.

The living primates provide us with a range of adaptive diversity that by analogy allows us to speculate about the adaptive nature of our ancestors. Thus, the combined field and laboratory studies enable us to learn more about man's relatives and enrich understanding of ourselves and of our origins.

Primatology has many practical applications. For example, a whole series of bio-medical questions have been dealt with and a variety of medicines developed through the study of captive primates. Extensive laboratory analysis of primates, especially monkeys, has been devoted to the study of nutrition, infectious diseases, deficiencies of the heart and circulatory system, arteriosclerosis, diabetes, and cancer. The production of new vaccines and the testing of these have been advanced greatly through the use of primate subjects. Not the least of their uses was the involvement of nonhuman primates in the space exploration program.

In modern times primatologists also raise their voices in warning that the wild populations of many primates are under imminent threat of extinction. Because the survival of the entire environment is a human responsibility, we must insure the future of all living organisms and not just our own kind. If lemurs, lorises, tarsiers, monkeys, and apes should become extinct, then we will certainly lose the chance to further understand the pathway through which we ourselves arose. Even though extinction is not an uncommon event, we alone can avoid being the cause of animal extinction.

Development of primatology in the last twenty years has not only produced new comparative insights but has shown as well that there are very many aspects of primate history and biology about which we know little. In consequence, beyond present understanding, broad perspectives open for future research. We see primatology today as a young, vigorous science. A distinct separation between ourselves and the most closely related non-human does not exist. Yet *Homo sapiens* stands as much more than an animal: Our species alone exhibits the ability to think and speak and the capacity to produce civilization, culture, science, and religion.

chapter 2
Definition of Order Primates

The Linnaean Order Primates, has no doubt stimulated more scholarly and popular interest than has any other major group of mammals. These creatures have a long history: They first appeared in the form of the late Cretaceous genus *Purgatorius*. An addition to the interest arising from the fact that we ourselves are ranked in this mammalian Order is the diversity of the group, which includes over fifty extant genera. This living diversity makes it the seventh most populous order of mammals in terms of generic groups; only Marsupialia, Insectivora, Rodentia, Carnivora, Chiroptera, and Artiodactyla have more genera. The great generic diversity among primates is emphasized by the high variety of locomotor and social systems, and both systems probably show greater variation within the order than can be found within any other major mammalian group. Adding to the entire series of living forms that provide a sequence of grades of organization roughly approximating a scale or ladder of nature are the great variety of fossil genera. There are about twice as many extinct genera of primates than there are living. The number is now over 105 and is rising every year with the description of new discoveries from the past.

Primates can be defined as placental mammals having orbits encircled with bone, with clavicles present, and with flat nails on at least some digits; the brain tends to be large relative to body size and shows a posterior lobe and calcarine fissure. Typically, there are two pectoral mammae and opposable innermost digits on the extremities, together with pendulous penis, scrotal testes, and large caecum. Cheek-teeth tend to be simple and low-crowned, often with secondary development of surface wrinkling and upper molar hypocones.

Order primates has two suborders: Prosimii, the prosimians or premonkeys, and Anthropoidea, anthropoids, or more correctly anthropoideans or higher primates, including Old and New World monkeys, apes, and humans.

Besides these two suborders of the mammalian Order Primates, there is another group, the tree shrews or tupaias, which were formerly ranked by most authorities in primates. A preponderance of present evidence would place them as a generalized side-branch of insectivores or as a separate order: either Tupaiaoidea or Scandentia. In spite of the trend to classify tupaias apart from Order Primates (see Hill, 1953; Van Valen, 1965), little has been done to justify that they are actually closer to other Insectivora, whose principal internal subdivisions have long been separate, at least since the Cretaceous Period. Even if tree shrews are considered as belonging to an order separate from primates, they do indicate fairly closely what we think the Cretaceous forebears of primates looked like. Hence, tree shrews are often kept and studied together with primate colonies by primatologists. For these reasons, tree shrews are included in this book, together with our survey of living prosimian primates. For those who wish to consider this question further, see Luckett et al. (1980). Despite extensive analysis by many authors, the question of whether or not to exclude the tupaiids from the order primates has not been definitely resolved. The three mammalian groups can be defined as follows:

1. *Tupaiiformes.* An infraorder (or order) resembling primates in the possession of a relatively large braincase, having eyesockets rimmed by a circle of bone and with males possessing a pendulous penis. Tupaiiformes differ from primates in lacking flat nails on any digits, all of which are clawed and the large toe aligned with the other digits; they also differ in having the bony floor of the middle ear made up from a different bone than the one that encloses the inner ear, and lastly, they have either premolar-like upper canines or none.

2. *Prosimii.* A suborder of primates differing from tree shrews and other non-primates because they have a petrosal bulla of the ear, typically a higher degree of orbital frontality, and flat nails on some or most of the digits. Hindlimbs are usually considerably longer

than forelimbs. Prosimii differ from Anthropoidea because they have no post-orbital closure, no fusion of metopic suture between frontal bones and the symphysal suture between the mandibles; these are sutures where closure either does not occur or appears late in individual development. Prosimii are also different because they typically have procumbent lower incisors (or tooth combs) and at least one toilet claw on the hindfoot.

3. Anthropoidea. A suborder of primates in which eyesockets are enclosed by bony plates and in which there is a fusion of the midline of the two halves of the mandible and in the forehead of the primitively dual frontal bones into single bones (mandible and frontal), and the auditory region is characterized by loss of the stapedial artery of the ear.

SURVEY OF THE LIVING PROSIMII AND THE TUPAIIFORMES: PART 1. TUPAIIDAE

As we have already seen, this group of relatively unspecialized arboreal placentals may well give us a general impression of what the earliest forerunners of the primates were like, and therefore they are of considerable interest to primatologists regardless of which order they are placed in. The lengthy debate (see Luckett et al., 1980) about whether these animals should be placed with Insectivora or with primates demonstrates the fact that they occupy a broad transitional position between certain members of these two orders. The intermediate anatomical characteristics of Tupaiidae certainly suggest that this type of mammal and the primates were interrelated in origin.

A first impression of the behavior of these small animals is one of tenseness and nervousness. They show considerable size range from species to species and their general body form differs from any prosimians. Tupaias look more like squirrels than like lemurs, except that their long and pointed snout resembles the latter. The family Tupaiidae contains two subfamilies, Tupaiinae and Ptilocercinae.

The subfamily Ptilocercinae contains one genus—*Ptilocercus*, with a single species—*P. lowii*. This monotypic species occurs in the tropical rain forests of North Borneo, the island of Bangka, the south Moluccas, and northern Sumatra as well as in nearby outlying islands. The species' common name is the "Pen-tailed Tree Shrew."

It is a rather small-bodied form (body and head length 12–15 cm, tail length 16–18 cm).[1] The forearms are relatively long, averaging about 4/5 of leg length. This figure for *Ptilocercus* averages 80, intermembral index that is determined thus:

$$\frac{\text{humerus} + \text{radius} \times 100.}{\text{femur} + \text{tibia}}$$

The tail is tufted in the third away from the body or "distal" third. It is sparsely haired in the "proximal" region, i.e., near the body. Moreover, the bare part is scaled—each scale followed by three short hairs. The large, mobile, and divergent ears are quite unlike those of the Tupaiinae in form and are relatively large for an animal of this size.

In many characteristics *Ptilocercus* stands even closer to a hypothetical basal placental than do the tupaiines, but apart from these primitive features, it is much like the tupaiines in appearance and behavior. "Pen-tails" are crepuscular (active at dawn and dusk), and nocturnal, and like tupaiines they inhabit the lower canopy of the forest. Their eyes are relatively larger and more frontally oriented than in Tupaiinae; the snout is somewhat shorter. These small mammals have been said to feed primarily on insects and also occasionally on small lizards. *Ptilocercus lowii* builds nests in tree holes or in the forks of branches.

The subfamily Tupaiinae contains three genera: *Tupaia*—the tree shrew (with between eleven and fourteen species), Figure 2–1, *Dendrogale*—the smooth-tailed tree shrew (with two species), and *Urogale*—the Philippine tree shrew (with one species). In *Tupaia* the body size ranges from 10–24 cm, and the tails are long, between 13 and 22 cm; in *Dendrogale*, the body size ranges from 10–15 cm, and the tails range from 10 to 15 cm; in *Urogale*, the largest tree shrew, body size ranges from 18–24 cm, and the tail ranges from 14 to 17 cm. These species live in the rain forests of Southeast Asia, in India north of the Ganges and south of the Himalayas, in Burma, Sumatra, Java, Borneo, Bali, and in many of the smaller East Indian Islands, with ranges sometimes extending up to about 3,400 meters above sea level. All species are diurnal and inhabit the lower forest canopy, bushes, scrub, and in some instances the ground. The tail in these animals is bushy and, as mentioned, the general appearance, except for the more pointed snout, is like that of squirrels. In fact, the Malay word *tupaia* means "squirrel." Their eyes are relatively

---

[1] Body size means head plus body length if not otherwise indicated.

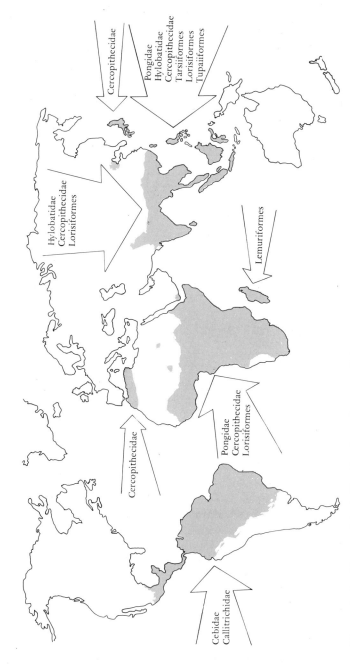

Figure 2–1 Geographical distribution of extant Primates.

large, and the snout has a glandular rhinarium or wet, sensory skin of smell, as in the dog and cat (see Figure 7–1). Their ears are immobile and they lie against the head, having a shape and general outline that rather closely resemble those of man (see Figure 2–2). In *Tupaia* the hindlimbs are longer than the forelimbs and the intermembral index falls around seventy-three, much lower than in *Ptilocercus*. All fingers and toes of tree shrews have pointed claws. Even so, tupaias appear to have greater dexterity in holding food items than do squirrels. The elongated and anteriorly narrowing skull of the tree shrew, its laterally directed but comparatively big eyes, its long, bushy tail, its nearly equal limb length (legs are somewhat longer than arms), and the rapidity of its nervous movements all add up to give the impression that these animals are more primitive than the true primates. Tree shrews seem to live in pairs and are known to have a high level of intraspecific aggression; they sleep in nests that they build in tree holes and other suitable places, and here the young are born and raised. In contrast to most primate species, tree shrews have multiple births or litters, and the young are left in a nest. Tupaias are most active during morning and early evening. Tree shrews occasionally eat seeds, fruit, and shoots

Figure 2–2 *Tupaia glis*. [Photo courtesy of Heinrich Sprankel.]

but more often insects, insect larvae, small lizards, birds and birds' eggs, and even snails; that is, they are omnivorous. In captivity, they accept meat, fruits, vegetables, eggs, insects, mice, lizards, and earthworms.

SURVEY OF THE LIVING PROSIMII AND THE TUPAIIFORMES: PART II. PROSIMII

Included here are at least four Madagascan families: Cheirogaleidae, Indriidae, Daubentoniidae, Lemuridae[1], and three families from Africa and Asia: Galagidae, Lorisidae, and Tarsiidae.

Compared to the other primate suborder Anthropoidea (including all the monkeys, apes, and men), Prosimii or lower primates are in general smaller, simpler, and less advanced animals. Relative to body size, the brain of all the prosimians is comparatively small, and many of them do not look like monkeys but rather more like primitive mammals. All twenty species of four of these prosimian families live on the island of Madagascar (Lemuridae, Indriidae, Cheirogaleidae, and Daubentoniidae) and in the Comoro Archipelago (two species of *Lemur* alone), where they are the dominant endemic mammals occupying many niches. No higher primates have ever reached Madagascar to compete with the Malagasy lemurs. Many of these Madagascan lower primates are diurnal and/or crepuscular. Species of the genera *Lepilemur, Cheirogaleus, Microcebus, Avahi, Phaner,* and *Daubentonia* are, however, nocturnal. Two other prosimian families (Lorisidae, Galagidae) have developed in Africa and one (Tarsiidae) in Eastern Asia, and all three include exclusively nocturnal or crepuscular animals. Lorisidae are also found in Asia.

All prosimians have catlike or rather pointed ratlike muzzles. Their eyesockets are rotated forward and nearer to each other than are those of Tupaiidae or most nonprimate mammals. In animals that have their eyes on either side of the head, the fields of vision are totally different and overlap little if at all. In living *Tupaia* the angle between the visual axes of both eyes is still 140 degrees, whereas among lower primates this angle is reduced to 70 or 60 degrees. Thus, in *Tupaia* partial binocular vision appears to be possible, but at the same time this entails a reduction of the total area seen at one

---

[1]Rumpler (1974) separates from Lemuridae the genus *Lepilemur* as a fifth family Lepilemuridae.

time. A binocular field of vision is the key to stereoscopic vision and is necessary for estimating distances in three-dimensional space. Having these forward-directed eyes is sometimes called "orbital frontality." This latter evolutionary achievement is highly important for a successful life in a three-dimensional arboreal habitat: the basic primate environment. Cartmill (1972), however, presented the case that frontality initially evolved mainly to facilitate precision in food catching and gathering.

Like all other primates, prosimians have highly grooved and ridged tactile skin on the plantar aspect of hands and feet. These are the dermatoglyphics: finger prints or dermal rugosities. Prosimians also have nails on the tips of most toes and all fingers except in the case of the aye aye, *Daubentonia*. This unique species has claws on fingers and toes except on the great toe or hallux, where it has a nail. All the other Malagasy lower primates have a claw only on the second toe (the so-called "toilet claw"). This claw pattern also holds for all the non-Malagasy lower primates except *Tarsius*, whose second and third toes are clawed for grooming. All the lower primates have a widely abducted first toe (hallux): this is one of the basic primate adaptations for grasping. The hands and feet of lower primates are typically five fingered. In all the lower primates except *Tarsius*, the area around the nose and lip is covered by a moist skin, the rhinarium. The rhinarium is a "close-up" sense organ found in most other nonprimate mammals. This organ is perhaps the most modified in galagos and mouse lemurs (*Microcebus*) and is lost among the higher primates and *Tarsius*. In prosimians the tail is reduced in the Malagasy genus *Indri* and in the African and Asian lorises (excepting *Galago*). All of the extant lower primates are basically arboreal; only one of the surviving lemurs, *Lemur catta*, has adopted (partly) terrestrial habits, but several possibly terrestrial subfossil lemurs are known.

Two subfamilies were once thought to belong in Lemuridae: the Lemurinae and the Cheirogaleinae. Rumpler (1974) revised the classification of Lemuroidea. He raised Cheirogaleinae to the rank of a family as Cheirogaleidae, with two subfamilies Cheirogaleinae and Phanerinae. Cheirogaleinae contains the genera *Microcebus*, *Cheirogaleus*, and *Allocebus*. Phanerinae has only one genus, *Phaner*. The one species of *Allocebus* is a little bit of a mystery because it is known only from two or three specimens. Within the Lemurinae there are four genera (Figure 2–3): *Varecia*, *Lemur*, *Hapalemur*, and *Lepilemur*.

The genus *Lemur* (six species) is often generally regarded as the "typical" lower primate. Species of this genus live in the tropical

Definition of Order Primates 17

Figure 2–3 *Lemur variegatus* or *Varecia*. [Photo courtesy of Michael D. Stuart and the Duke University Primate Center.]

rain forest and the drier seasonal forests in Madagascar and some of the Comoro Islands and seem to prefer to live and move among the large horizontal branches of the forest. *Lemur catta* differs in that it spends much of its active time on the ground. Lemurs have foxlike heads with big, sometimes tufted, ears, dense fur and long, bushy tails (Figure 2–4). Their body length ranges between 30–60 cm. Many lemurs have eye-catching contrasting fur colors, and these

Figure 2-4 *Lemur fulvus rufus.* [Photo courtesy of Michael D. Stuart.]

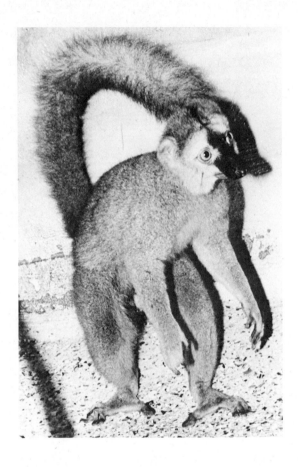

may vary intraspecifically. *Lemur macaco* males are jet black, the females rusty red-brown, a case of sexual dimorphism of the fur color (sexual dichromatism). Newborn *Lemur macaco* males and females alike are dark. Although ear-tufts of females are white, the fur of the female turns lighter after only a few weeks. The tail in *Lemur catta* has conspicuous black and white alternating rings; it is typically held upright in an S-shaped curve when the animal is moving quadrupedally. This highly conspicuous tail of *Lemur catta* functions as a social signal (Figure 2-5). Many of the lemurs have ruffs around their necks that are also identification markers. (Figure 2-6). With lemurs, bushy tails are always longer than the body plus head together; the hindlimbs are longer than the forelimbs with intermembral index around 68-70. The eyes of these animals show many bright colors with yellow, green, blue, or orange irises; their glances include frequent inquisitive staring and, like higher primates, they can intimidate by staring. Their ability to grasp small objects is not very highly developed. Rather, they rake up objects between the tips of the fingers—two (index) to five—and the palm, the thumb being widely abductable but not typically brought into real opposition to fingers 2–5. Brown, black, and ring-tailed lemurs are gregarious animals that live together in groups of adults of both sexes with young animals of different ages. Throughout the lemurs, social groups vary—couples, multi-male, single male, and open groups where males may come and go. Red-bellied and mongoose lemurs appear to be pair bonded as are ruffed lemurs, two adults traveling with their young. Some groups are reported to have well-defined territories; Lemur marking behavior is probably related to this. *L. catta* have cutaneous glands for scent marking located on their forearms; urine marking also occurs. The tail can be used to spread the olfactory messages as can the hands, head, and mouth.

Figure 2–5 *Lemur catta.* [Photo courtesy of Christian R. Schmidt.]

Figure 2–6 Ruff around the neck of *Lemur fulvus albifrons*. [Photo courtesy of Michael D. Stuart and the Duke University Primate Center.]

Interestingly, *Lemur catta* can often be observed sunning in an upright sitting position, leaning back on a vertical support with arms outstretched in almost manlike posture. Species of the *Lemur* have an extended polyestrous breeding season with births occurring toward the end of the austral winter (usually varying from September to November). During the May to August mating period, increase in marking activities can be observed. The diet of most lemurids consists mainly of a combination of fruits, flowers, flower buds, nectar, and leaves.

Varecia—the ruffed lemur—was long classified as a species of genus *Lemur*. The one extant ruffed lemur species inhabits an extended north–south range in the high forest canopy of the central eastern rain forest. Body size is slightly greater than any of the species of genus *Lemur*. Like *Hapalemur* species, *Varecia variegata* associates in pairs or nuclear families from which the young are ejected when they reach sexual maturity. Thus, groups seldom exceed four or five individuals. This is the only large daytime active primate in which litter size averages two or above, and triplets are common. Newborns cannot climb or cling and are kept in large loosely constructed nests. Ruffed lemurs emit loud barking cries in unison that have been termed "mob calls." The males scent mark with throat glands. Ruffed lemurs breed well in captivity, and behavior and reproductive studies are based on captive colonies. This species has never been studied in the wild, where it is increasingly rare.

Lepilemur—the sportive lemur—has many species and is widely distributed in the forests of Madagascar and on the island of Nosy Be. *Lepilemur* species have a body length of 28–36 cm, the tail is between 25–28 cm in length. Lepilemurs have short muzzles, big eyes, and rounded ears. The hindlimbs, although shorter than in other lemurids, are considerably longer than the forelimbs (intermembral index averages around 60). The sportive lemurs usually jump in a vertical position, leaping from one upright support to another. They are adapted to many habitats, including both rain forests and zerophytic scrub forests. Mainland sportive lemurs usually sleep rolled up into a ball in tree holes, whereas those on Nosy Be sleep on exposed thick branches. The difference in sleeping habits seems to be correlated with the total absence of predators on Nosy Be. Sportive lemurs are nocturnal animals that mostly live solitarily but seem, however, to be concentrated in greater density in certain areas. The breeding season is apparently restricted to the period from May through August. Mothers may carry their youngsters in the mouth. Marking behavior is seldom seen despite the fact that the males have scent glands in the scrotal skin. Sportive lemurs have a variety of loud, mainly territorial, calls. Their diet consists of leaves, flowers, bark, and some fruit.

Hapalemur—the gentle lemur—has two species: *H. griseus* and *H. simus*. The latter species, however, is highly distinctive, perhaps even a separate genus. Some recent authorities had even believed it to be extinct, but we now know that *H. simus* still survives in a restricted spot in eastern Madagascar. Species of *Hapalemur* live in the lower arboreal stratum on large, more or less horizontal branches, but also move about in lower small trees and bamboo thickets. Their body length ranges for the genus from 26–27 cm, the length of the bushy tail between 24–37 cm. The intermembral index averages about 67. The face of both *Hapalemur* species is less elongated and tapering than that of genus *Lemur*. The ears are hairy, and the eyes appear to be closer together than in the latter. Gentle lemurs seem to have well-defined territories and live in small groups of between three and six animals. They have forearm and axial glands used for scent marking purposes. Gentle lemurs are seldom bred in captivity.

Two species of dwarf lemurs—*Cheirogaleus major* and *C. medius*—live in the lower strata of foliage in eastern, western, and southern forests of Madagascar. A third related species—*Allocebus trichotis*—is known only from two or three individual specimens. Dwarf lemurs are nocturnal and are smaller than gentle lemurs (19–27 cm body length, 16–17 cm tail length); the head appears globular when compared with the head of genus *Lemur*, but the muzzle is somewhat more pointed than that of *Hapalemur* species. The ears are relatively small and are membranous, not haired. These animals are all nocturnal, with comparatively large eyes. Dwarf lemurs are quadrupedal. Their hindlimbs are, of course, longer than the forelimbs but not as elongated as in *Lepilemur* or the indrids. The intermembral index of *Cheirogaleus* averages around 71. The grip of the hand in these small lemurs resembles that of the South American monkeys. Objects are taken up and small branches are grasped between the second and third finger, not between index finger and thumb. The nails of cheirogalids are somewhat keeled and pointed. The dwarf lemur has periods of inactivity or torpor. The tail in dwarf lemurs serves as an area for storage of fat that fluctuates seasonally, most of the fat being absorbed during the period of greatest food scarcity. Species of this genus are reported to live solitarily or in pairs, as seems to be a special case for all strictly nocturnal prosimians. Fecal and scent marking activity of these lemurs is important. Dwarf lemurs nest in tree holes and often sleep together in curled up positions; their diet evidently consists primarily of fruits, flowers, and perhaps gum.

The genus *Phaner* is found widely throughout the western rain forest of Madagascar and has but one species. The fork-marked Dwarf Lemur, *Phaner furcifer*, is somewhat larger than the lesser dwarf lemur, the smaller species of genus *Cheirogaleus*, *C. medius*. The body length plus head length is 25–28 cm, the length of the tail is 32–35 cm. This species shows a dark ring around each eye, which continues backward in converging black stripes up to the top of the head where the lines fuse into a single black median stripe over the back. These Y-shaped markings give the animal its common name. In this species the snout is relatively blunt, the eyes large, and the animal nocturnal. With *Phaner*, the ears are considerably larger than those of the species of *Cheirogaleus* and differ also because they are more pointed. In *Phaner* the nails are more strongly keeled and pointed than in *Cheirogaleus*, and are comparable to the nails of *Euoticus elegantulus*, the needle-clawed bush baby (Petter et al. 1971, 1977). These animals have been reported to live solitarily, but larger groups sometimes congregate, whether for feeding or breeding is not clear. *Phaner* moves quadrupedally and leaps more often than *Cheirogaleus*. The intermembral index is lower than in dwarf lemurs, around 67. *Phaner* eats gum and eagerly licks syrup-like sticky liquids of insect larvae (Homoptera and Coleoptera).

Microcebus and *Mirza*—the mouse lemurs—are now recognized as two separate genera (Tattersall, 1982). Mouse lemurs are abundant in all forests of the east and west coast of Madagascar. One of the two genera—*Microcebus*, the "lesser mouse lemur" (Figure 2–7)—together with the one species of *Cebuella*—the pygmy marmoset—and *Galago demidovii*, are the smallest of primates. *M. murinus* and *Cebuella* both have a body length of only about 13 cm, but the tail of *Micocebus murinus* is shorter than that of *Cebuella*. *Mirza coquereli* on the other hand, is about the size of a common American grey squirrel or species of the genus *Phaner*, with a body length of around 25 cm. The tail of *M. coquereli* is slightly shorter than in *Phaner furcifer*, measuring only 28 cm as compared to 33–

Figure 2–7 *Microcebus murinus*. [Photo courtesy of Michael D. Stuart and the Duke University Primate Center.]

35 cm. Genera *Microcebus* and *Mirza* are nocturnal, with the commensurate large eyes. Mouse lemurs have a globular head and a short but pointed muzzle. *M. murinus* is a grey-colored animal, but there is a red form with smaller ears in the eastern forest that is sometimes recognized as a separate species, *M. rufus*. The mobile ears of mouse lemurs are membranous and rounded. Hindlimbs are just about as short relatively as those of dwarf lemurs, with a similar intermembral index of around 71. The heel bones of the foot are slightly elongated. Mouse lemurs run quadrupedally in a mouselike manner. They can also leap with agility, as do many small primates. Mainly, mouse lemur males live alone, and it has been reported that females sleep in groups in nests in tree holes or in bushes. Males show marking behavior before copulation. Both mouse lemur genera have a defined breeding season during the Madagascan spring and summer (September to December). Mouse lemur newborn babies are always transported in the mother's mouth. Like dwarf lemurs, mouse lemurs are said to go through periods of torpor or inactivity that are, however, less distinct and shorter than in *Cheirogaleus*. As with the Dwarf lemurs, mouse lemurs show seasonal storage of fat in the tail (as do many mammals that hibernate), but here fat storage may offset food shortages. The diet of these small animals consists primarily of fruit and insects.

Genus *Avahi*—the smallest member of the Indriidae—has only one species, *A. laniger*. Avahis, called woolly indris, or footsy-fe, meaning "white-thighs," live in the rain forests of estern Madagascar and in dry forests of the northwest of the island. The length of the bodies varies between 30–33 cm, and the bushy tails are considerably longer, about 39–40 cm. These noctural animals have dense and soft fur. The head is round, the muzzle blunt, and the ears are not very conspicuous because they are hidden in the fur. As is typical for nocturnal animals, the eyes are comparatively large. The hindlimbs of avahis are considerably longer than the forelimbs: intermembral index around 56. The third, fourth, and fifth digits on hands and feet are linked together by a skinfold (syndactyly). Avahis move about by clinging on vertical supports and by leaping great distances. They are reported to sleep rolled up into a ball in the outer branches of trees during the day. These rare and little-known creatures are said to live in small family units consisting of a breeding pair and one or two offspring. The mating season seems to be restricted, and newborn avahis are seen only during late August and early September. The babies ride on the mother's belly and later on her back. No marking behavior has been reported in spite of the presence of cutaneous glands on the scrotum of male animals.

Definition of Order Primates

Unlike the *Indri*, avahis are quiet and do not employ noisy vocalizations for territorial purposes. The diet of avahis is predominantly leaves, buds, and bark of a few trees only, although they may also eat some ripe fruit. Avahis have never been kept successfully in captivity.

Sifakas—genus *Propithecus*—are found over most forested areas of Madagascar except the far northeast of the island. The two species of the genus *Propithecus*—*P. diadema* and *P. verreauxi*—rarely come to the ground in the natural habitat. The body length is from 45–54 cm, and the long tails measure between 48–56 cm. The head is round, and the snout is not very prominent. Both species have black faces and black ears. The woolly fur is colorful, the colors extremely variable. Diademed sifakas may be golden or black, whereas verreaux's may have maroon on shoulders or head. The eyes are large for diurnal animals and are directed forward, giving a

Figure 2–8 *Propithecus verreauxi.* [Friderun Ankel-Simons photo. Courtesy of the Duke University Primate Center.]

staring impression. The irises of the eyes are usually bright golden-yellow but may be blue-green or variously colored. The hindlimbs of sifakas are much longer than the forelimbs, with an average intermembral index around 60. The native names for *P. verreauxi* of "sifak" or "tsibuhak," and "simpoon" for *P. diadema*, imitate certain of their cries. These animals usually jump about in mainly vertical positions, often springing between vertical supports. Thus, they too are often placed in locomotor classifications as vertical leaping and clinging animals (Figure 2–9). Diademed and verreaux's sifakas live in groups that consist of three to nine animals. The core of these groups is usually a family unit. Newborn babies cling tightly to their mothers. The birth season is in July and August. The mating season seems to be strictly restricted to a few days about five months earlier. Male sifakas have cutaneous glands on the throat area and have been observed to throatmark. Female sifakas seem to mark primarily with urine and do not have the throat glands (Petter, 1962; Jolly, 1967; Richard, 1974). Like avahis, sifakas eat leaves, buds, bark, and, in lesser quantities, ripe fruit.

The name-bearing genus and species of the family Indriidae—*Indri indri*—have been well analyzed by Pollock (1977). The strange and exotic *Indri*—or "babacoto," pronounced "babakut"—inhabits the central eastern rain forest of Madagascar. The Malagasy people whose villages are near where the indri live regard them as "men" and their killing as "murder." These are the largest extant prosimians and the largest living primates of Madagascar, having a body length of about 70 cm. Moreover, the indri is the only Malagasy primate with an almost completely reduced tail (tail length 3 cm). The fur of the indri or "babakut" is silky and dense and may vary in color but is normally black and white. These animals may weigh 10–15 kilograms. Although large, they are strictly arboreal, moving around in the same way as do avahis and sifakas. Hindlimbs are pronouncedly longer than forelimbs, with an intermembral index averaging around 62. The head is globular, the snout slightly more pointed than that of the sifakas; their eyes are also comparatively large for an animal with entirely diurnal activity periods. First activities are around sunrise, but they stop feeding and moving about at about 3:00 in the afternoon. As with the avahis, the digits 3–5 of hands and feet are webbed (syndactyly). Indrises or indri live in family units of two to four individuals. Pollock has shown that the infrequent very loud calls strictly define the territories of each family of babakut. These are surely the noisiest of all the Malagasy lemurs. Their song can carry for two or three miles so that many travelers have heard the far reaching songs of these animals without

Figure 2–9 *Propithecus verreauxi*, in the madagascan environment. [Photos courtesy of Alison Richard.]

ever seeing them. In fact, the name indri means "look there (it is)" in the Merina dialect of the highlands. Indrises have a large laryngeal sac in the throat, presumably developed to enhance their vocal abilities. The indri feeds on foliage, flowers, fruit, and leaf buds, mainly from tree laurels (Pollock, 1977).

Daubentonia—a genus and family with one species—is perhaps the most peculiar of the Malagasy lemurs. The "aye-aye," now very rare, seems to be restricted primarily to the eastern forests of Madagascar. The fur is coarse, long haired, and shaggy, unlike that of all the other Malagasy lemurs. The tail is long and bushy, very thick with long fur. Aye-ayes have a body length measuring about 40 cm, whereas the tail length is between 55–60 cm. Unlike other lemurs, *Daubentonia* has claws on hands and feet except on the great toe. It is thought that these "claws" are a secondary development from nails. The third digit of the hand is thin and elongated and is used by the animal to "fish" beetle grubs and other insects out of cracks. The hands and fingers of *Daubentonia* are very large and long, compared to all other prosimians, and fingers and toes are almost the same length. The dentition of *Daubentonia* is unique among extant primates; it is highly specialized and anteriorly resembles that of rodents. The muzzle is comparatively

Figure 2–10 *Daubentonia*, from Owen 1866.

high and narrow as a result of the narrow front teeth. Perhaps in correlation with enhanced manual dexterity, the brain is relatively larger and more convoluted than in other lemurs. The eyes of the aye-aye are big and directed slightly upward and forward. Enormous membranous ears add to the peculiar appearance of the broad head. All four limbs are comparatively short, the hindlimbs longer than the forelimbs: intermembral index averages 71. The activity pattern of this species is noctural and solitary. They build and sleep in large nests. Their diet includes fruits and insect larvae. Villagers hold them in dread and will move away from places where they have been seen (Figure 2–10).

Family Lorisidae

The genus that gives its name to the family Lorisidae has one species: the slender loris, *Loris tardigradus*. Slender lorises are restricted in distribution to southern India and Ceylon. Their body length varies between 18–27 cm and the tail is very short. Fore- and hindlimbs are nearly equal in length and elongated: intermembral index averages 91. Fur is shorter and sparser than in the other Asian

Figure 2–11 *Loris tardigradus.* [Photo courtesy of Dieter Glaser.]

lorises, thus giving the impression of slender limbs. Although the fur is not very dense, their heels are haired. Lorises usually climb slowly, as *tardigradus* implies, typically holding on with three limbs at a time. Although noted for their slow movements, they can move very quickly if necessary, as can the three other genera of the Lorisidae (Walker, 1969). Slender lorises are arboreal and inhabit tropical rain forests. Individual lorises are said to live solitarily. During the day they sleep rolled up in crotches between the branches of trees, securing their position with the clasping feet. Lorises appear to have two mating seasons a year, in April–May and in November. Slender lorises usually have single births, but twin births can occur. Urine marking and urine washing are intensively employed by lorises. These prosimians are active during the night (as are all members of Lorisidae) and feed mainly on insects, small birds, and lizards.

The common name of the genus *Nycticebus* is the slow loris. *Nycticebus*, with two species, has a wide distribution over the Southeast Asian continent and islands. The animals inhabit the high strata of the tropical rain forests. Slow lorises have a comparatively more bulky appearance than the slender lorises. The head is round; the snout appears less pointed than in slender lorises; the ears are membranous, round and comparatively small and are hidden in the dense fur. The area around the eyes, especially above them, is dark furred (Figure 2–12). The dark area, however, is less circumscribed than in slender lorises. The common species of *Nycticebus*—*N. coucang*—has a body length of between 27–38 cm. The outer tail is totally reduced: only a short tail stump is present and disappears in the dense fur. The other species—*N. pygmaeus*—is smaller than *N. coucang*. In *Nycticebus* the second digit of the hands is somewhat reduced. *Nycticebus* has powerful hands and especially powerful feet, as compared to those of the slender loris. The limbs are nearly equal in length, with an intermembral index averaging around 89. No distinct breeding season has been recorded for slow lorises. Slow lorises live singly or in pairs. Nothing is known about territoriality of *Nycticebus* in the natural habitat; they do, however, mark cages with urine. Slow lorises have been observed to groom themselves very carefully. A wide variety of vegetation and some animal foods combine to make up their diets.

The African genera of Lorisidae can also be divided into two types, one being more slender, the other more like the slow loris of Asia. The slender African loris is called "the angwantibo" or "golden potto." Its generic name is *Arctocebus*, with a single

Figure 2-12 *Nycticebus coucang.* [Photo courtesy of Michael D. Stuart.]

species, *A. calabarensis*. These animals are restricted to the tropical forest of West Africa, where they live in the lower forest strata. Angwantibos are slightly larger than slender lorises, their body size between 22–27 cm. The tail is absent. The limbs of *Arctocebus* are thin, the body is cylindrical and has a "blunt" backside, unlike that of any other prosimian. Hands and feet are dissimilar to most other primates. The index finger is reduced to two short phalanges, and the great toe is positioned opposite digits 3–5 in the hand. Digits 3–5 are webbed by a skin duplication up to the articulation between the proximal and the middle phalanges. Digits 3–5 of the foot are webbed to the same extent as the digits of the hand. The heels are haired. The head of *Arctocebus* somewhat resembles that of the slender lorises, globular with a pointed muzzle. Its ears are rounded and not very large if compared to those of the slender lorises. The eyes are larger than in diurnal prosimians but smaller than those of the slow loris. The locomotion of *Arctocebus* is less slow or deliberate than that of slow lorises and pottos; it seems that they do not always grasp a support with three digits of the hands and feet but may do so with only two of them at a time. Their activity period is primarily at night, but they have also been observed moving about during the day. Their diet consists of approximately 85 percent insects and 15 percent fruit (Charles-Dominique, 1971).

Pottos—*Perodicticus potto*—inhabit the same area as angwantibos, in the tropical forests of West Africa. In addition, their distribution extends further east than that of *Arctocebus*, into the western areas of Uganda and Kenya. In Gabon and Cameroon, where they have been observed in the same area as the angwantibo, they are reported to differ from the latter in preferring the upper stratum of the tropical forest. The body size of pottos varies between 33–42 cm. The tail is reduced, but a short external tail of about 6 cm in length is always visible, varying in size from individual to individual. The overall appearance of pottos is close to that of slow lorises. They are stout and robust in body form and in the limbs: both look like tiny bears. Fore- and hindlimbs of pottos are of nearly equal length, the hindlimbs somewhat longer, with an intermem-

Figure 2–13
Perodicticus potto.
[Friderun Ankel-Simons photo. Courtesy of the Duke University Primate Center.]

bral index around 88. The hands and feet of pottos are powerfully adapted for grasping. Apparently to extend the size of the grasp of other digits, the index finger is nearly totally reduced. Thus, the thumb is positioned opposite digits three to five of the hand. The hands and feet act like pairs of grasping tongs. All digits are very broad. The heel is haired. The head appears globular but slightly flat, with relatively small, rounded ears and big eyes. The eyes of *Perodicticus* are, however, relatively smaller than those of *Nycticebus*. In *Perodicticus* the dorsal spines of the lower neck and upper breast vertebrae are very much elongated and pointed at their tips. In pottos the spines even protrude nearly to the surface of the skin. They are covered by a cornified epithelium and surrounded by dense fur. The covering skin is highly sensitive. This sensitivity of the skin of the dorsal neck region is important in the social behavior of the pottos, as has been demonstrated by Walker (1970). Pottos are highly territorial in their natural habitat and live in small family units. Both urine marking and male scent-marking with circumgenital glands have been reported (Charles-Dominique, 1978). The diet of pottos consists principally of fruit and gum; they also eat some insects.

The genus *Galago*—the bush babies—is split into three subgenera: *Galago, Euoticus,* and *Galagoides*. Subgenus *Galago* (three species) is widespread over Africa south of the Sahara, except the far south. Species of the subgenera *Euoticus* (one species) and *Galagoides* (one species) are much more restricted to equatorial western and central Africa. *Euoticus* is primarily restricted to the tropical rain forests of Spanish Guinea.

Galagos live in both the tree savannah and the forest areas and are arboreal animals. Bush babies have globular heads, pointed muzzles, and very large, membranous ears. These ears are not covered with hair and are relatively the largest ears of all primates. Their eyes are very large and forward directed. Bush baby digits are comparatively slender and enlarge to circular terminal plantar pads on the tips. Subgenus *Euoticus* is also called the "needle-clawed galago": the nails of *Euoticus* are elevated medially and pointed at their tips. The largest of the galagos (*G. crassicaudatus*), the thick-tailed or great-tailed bush baby, measures between 30–38 cm, the tail is 41–47 cm long and is thick (Figure 2–14). The smallest of the galagos is (*G. (Galagoides) demidovii*), the dwarf galago. Its body length is only 12–16 cm, tail length 18–20 cm. The dwarf galago lives only in the lower, bushy forest stratum. Galagos often both rest and leap in a vertical position. They leap with ease, and they are rapidly propelled by the powerful hindlimbs that are considerably

Figure 2-14 *Galago crassicaudatus*, with cross-folds of the ear. [Photo courtesy of Michael D. Stuart.]

longer than the forelimbs, intermembral indices range from 57–64. Occasionally these animals move quadrupedally. This condition is especially true for dwarf galagos that often run quadrupedally along small branches. On the ground, galagos hop bipedally. Bush babies build nests for sleeping and pregnant females establish separate nests to give birth. The females maintain the nests with the offspring for some time after parturition. *G. crassicaudatus* has a breeding season in August and September. In the Sudan newborn bush babies are seen in April and September. Mothers often transport infants in their mouths, as Sauer (1967) described. There is a great deal of variation in the diets of galagos. *Galago alleni* and *Galago crassicaudatus* are primarily fruit eaters; *Galago demidovii* and *Galago senegalensis* are primarily insect eaters; and *Galago elegantulus* is primarily a gum eater.

Family Tarsiidae

The most exotic and the smallest major group of prosimians is the genus *Tarsius*. Tarsiers live on many islands off Southeast Asia. Three species of tarsier are distinguished: *T. syrichta* lives on some

Philippine islands and on Mindanao; *T. bancanus* lives in southeast Sumatra, Borneo, Banka, and also on other adjacent islands; *T. spectrum* lives on Celebes and some small surrounding islands. Their body size ranges from 9–16 cm, the tail length from 13–28 cm. Only the last third or half of the tail is hairy. These tail hairs are considerably longer than the short fine hairs of the dense fur that covers the body. The eyes of tarsiers are relatively the largest of all the primate eyes: a single eyeball of a tarsier has a greater volume than does the same animal's brain (Sprankel, 1965). Tarsiers have short trunks and comparatively big, globular heads, short, small snouts, and no wet rhinarium. The ears are membranous, rounded, and mobile. The name *Tarsius* is derived from the elongation of the tarsus or heel bones that, among all the primates, are the longest in this genus (Figure 2–15). *T. bancanus* has a special area of "friction pad" ventroproximally on the tail: an area used by *T. bancanus* as an additional support when clinging vertically. Suckerlike prehensile

Figure 2–15 *Tarsius bancanus,* young animal in a vertical clinging position. [Photo courtesy of Heinrich Sprankel.]

pads at the tips of the digits are characteristic of tarsiers. The second and third digits of the foot have claws. Tarsiers are the only prosimians with two grooming claws. Their locomotion is very specialized; tarsiers leap and cling, preferably on vertical supports. The intermembral index is low, about 56. Many tarsiers live in bamboo thickets, a perfect environment that provides vertical supports for an animal of a tarsier's body size. Tarsiers are, however, also found in the lower bushlike stratum of the tropical rain forest. Tarsiers are said to sleep sometimes clinging in a vertical position. Tarsiers live in pairs with one or two young of different ages, although females are sometimes seen alone with their offspring. As far as is known, tarsiers do not have defined breeding seasons. These small animals have been observed carrying their young in the mouth. Tarsiers are reported to be silent animals except for the very young and during mating, when both partners vocalize. Tarsiers establish and defend well-circumscribed territories. They show frequent marking behavior with their circumanal gland and by means of "urine-washing" with their feet and other common urine marking behavior. Grooming is rarely a social activity but mainly self-directed: tarsiers lick and rub their own fur and scratch themselves often with the two toilet claws of the feet. The diet of tarsiers consists mainly of animal protein: they live on insects and their larvae, small lizards, and nestlings of birds and other small prey.

For additional information on all aspects of prosimian biology, suggested readings include Martin, Doyle, and Walker (1974), Doyle and Martin (1979), Charles-Dominique (1977), and Charles-Dominique et al. (1980).

Suborder Anthropoidea

This suborder includes all monkeys, apes, and ourselves. The suborder is subdivided into three superfamilies: Ceboidea, Cercopithecoidea, and Hominoidea. The entire group of anthropoid or simian primates appears to be more homogeneous than does the suborder Prosimii. These anthropoideans are commonly called simian primates, from an old name for the suborder *Simii*; they can be subdivided into two infraorders. The division is in accordance with their geographical separation into New and Old World higher primates. Additionally, it correlates with a morphological difference of the face that can easily be recognized: all South American monkeys have flaring cartilaginous nasal wings and broad septa whereas in Old World monkeys, apes, and humans, the wings are positioned close to each other and this septum is narrow. The nares of Old World monkeys thus open downward and are placed near each other whereas those of the New World monkeys are directed upward and laterally. From the Greek words *platys* and *cata*, meaning broad and downward respectively, these two primate infraorders are named *Platyrrhini* and *Catarrhini*, broad-nosed primates and down-nosed primates. However, characteristics such as the cartilaginous nasal wings and septum are variable and consequently not of high taxonomic value. The terms *platyrrhine* and *catarrhine* are widely used in primatology since they usefully distinguish between the higher primates of the New and Old Worlds.

Many discussions have concentrated on the problems of primate taxonomy. Thus it has, for example, been questioned whether it is justifiable to subdivide the suborder Anthropoidea into the superfamily Ceboidea on the one side and the superfamilies Cercopithecoidea and Hominoidea on the other. These three superfamilies are groups of very different diversification, and consequently many different ways of dividing them up have been proposed. The same problem arises from a comparison of the two suborders Prosimii and Anthropoidea. Prosimii exhibits a great many more different adaptive types than do the Anthropoidea. It is becoming increasingly clear that a group covering Platyrrhini together with Catarrhini cannot be monophyletic.

Fiedler based his 1956 subdivision of the order Primates on Haeckel's (1866) suborders Prosimiae and Simiae. The correlative terms used by Simpson (1945) are the *Prosimii* (including the superfamilies Tupaiiformes, Lemuriformes, Lorisiformes, and Tarsiiformes) and the *Anthropoidea*, containing the superfamilies Ceboidea, Cercopithecoidea, and Hominoidea. Most European (continental) scientists working with primates have generally agreed to follow Fiedler's taxonomy. This is not, however, the case for English-speaking scholars. They usually follow Simpson's 1945 classification and, as stated, subdivide the orders into the suborders Prosimii (Illiger, 1811) and Anthropoidea (Mivart, 1873). As mentioned before, many do not include the Tupaiiformes in the order. In this discussion we will follow the classification of Simpson as modified by Simons (1972).

Ceboidea

The New World monkeys—infraorder Platyrrhini—have recently been reclassified by Hershkovitz (1977). In this classification, the

Platyrrhini has one superfamily, Ceboidea, that consists of three families: Cebidae, Callimiconidae, and Callitrichidae. Cebidae is further divided into ten subfamilies, seven of which include extant genera. These seven subfamilies and contained extant genera are:

Aotinae
  Aotus
Callicebinae
  Callicebus
Pithecinae
  Cacajao
  Pithecia
  Chiropotes
Alouattinae
  Alouatta
Cebinae
  Cebus
Saimirinae
  Saimiri
Atelinae
  Ateles
  Brachyteles
  Lagothrix

The genus *Aotus* has only one species, *A. trivirgatus*, the night monkey or douroucouli. This is the only monkey that is nocturnal. These monkeys are also often called owl monkeys. Night monkeys live in the Amazon basin, and the area of their occurrence is limited by the Orinoco River in the north and the Andes in the west. Night monkeys are found as far south as the Gran Chaco. The head of these middle-sized monkeys (24–27 cm body length, 22–42 cm tail length) is dominated by comparatively large eyes that are typical of nocturnal animals. The head is globular, the face short, and the nose, of course, platyrrhine. The ears are nearly totally hidden in the fur. The bushy tail is usually longer than the body. The forelimbs are short relative to hindlimbs: intermembral index averages about 74. Douroucoulis run and leap quadrupedally. Pairs with their offspring live in territories and vocalize with a variety of different calls. The vocalizations seem to be correlated with *both* the territoriality and the nocturnal habits of these monkeys. Their "urine washing" behavior is also possibly correlated with territoriality, as in many prosimians. Night monkeys have a glandular area on the underside of the base of the tail. Night monkeys usually have single births. There is no evidence of a restricted breeding season. The male grooms the female only in the context of sexual activity. Very young animals cling to their mothers. Later, the offspring are carried most of the time by the male and possibly also by older siblings. After reaching this age, the offspring return to the mother only to be nursed. During the day, these monkeys sleep in tree holes of the forest canopy where they live. The diet of night monkeys consists of a combination of fruit, insects, small mammals, and possibly birds.

Genus *Callicebus*—titis, with three species *C. torquatus, C. moloch,* and *C. personatus*—falls approximately in the same size range as the night monkey (body size 28–39 cm, tail length 33–49 cm). These monkeys also occur in the Amazon basin, but in the northeast they do not cross the Rio Negro and Amazon. The western limits are the mountain ranges of the Andes. *Callicebus* occur throughout Brazil and range into Paraguay. They prefer low thickets and forest areas near rivers and perhaps are the American vicars of African swamp monkeys. The head is globular as in all the New World monkeys of small body size. The snout is not prominent, the nares are far apart, and the ears are hidden in the coarse fur. The face is only sparsely covered with short hairs. In *Callicebus* species the tail is longer than the body. The hindlimbs are long relative to the forelimbs, with an intermembral index of 73, similar to *Aotus* and to other Platyrrhini of comparable body size. These animals run and leap in a quadrupedal manner and are very agile. *Callicebus* species live in family groups in territories that are relatively small, well defined, and noisily defended against intruders. Titis vocalize frequently, and their chorus reaches far. The morphology of the mandibular angle and an enlargement of the hyoid bone are in accordance with the vocal abilities of the titis, and both morphology and function resemble the howling monkey in this part of the anatomy. The pair-bond in *Callicebus,* is said to last over years. A mated pair or nuclear family spends much time together comprising a family group; parents have up to three successive generations of offspring. *Callicebus* pairs frequently sit leaning on each other with their tails curled together. Young animals are carried by the father or an older sibling, except when they change over to the mother to be nursed. Social grooming within the family group is frequent, and copulation occurs comparatively often during the breeding season. Groups "fight" vocally with loud, prolonged "songs" and with threatening movements when they meet. These are primarily diurnal animals. *Callicebus* eat insects, small birds and their eggs, possibly small mammals, some leaves, and fruit.

The following three genera are considered a separate subfamily Pithecinae by many scholars; the name giving genus for this family, *Pithecia*, has two species, *P. pithecia* and *P. monachus*. The common name for these medium-sized monkeys is "Saki"; their geographical range is defined by the Amazon and Orinoco in the north, but they occur south of these rivers as far as the tropical rain forest extends. The head is round and the face short. The nasal openings are very far apart; the face is only sparsely haired. Sakis have long fur and thick bushy tails. The hair on the top of the head looks as if it has been combed in the direction of the face and sides from the center of the head, giving the anthropomorphic impression of a monk's hood. This hair pattern gives the common name "monk saki" to *P. monachus*. Saki ears are barely visible; they are hidden in the long hair. The male of *P. pithecia* has a broad band of white fur around the naked dark face. The body color of both female and male sakis is dark-brownish-gray. Their body size varies between 30–48 cm, the length of tail between 25–55 cm. The hindlimbs are much longer than the forelimbs with an intermembral index of 76. *Pithecia* species reportedly move primarily by leaping. All species of the three genera of this family, *Pithecia*, *Chiropotes*, and *Cacajao* share a peculiarity of the dentition: their lower and upper incisors are not vertically implanted in the jaws but are tilted into an extremely procumbent forward position. Thus, the incisors mainly function as a pair of pincers. Sakis have glandular areas on the throat of unknown function. *Pithecia* is one of the South American monkeys only rarely caught or observed and is seldom seen in zoos. Sakis live in family groups. Very little is known about territoriality, breeding, or social interactions in species of this genus. The diet reportedly is mainly vegetarian, with fruit and berries predominating.

Genus *Chiropotes*—the bearded sakis—has two species, *C. satanas* and *C. albinasus*. Some authors do not separate this genus from *Pithecia*. The geographic range of the bearded sakis appears to center along the Amazon River. *C. albinasus* (with a light-colored nose, the rest of the body dark) has been reported to appear as far south as the Gran Chaco. They are called "bearded sakis" because the beards of *Chiropotes* monkeys strongly contrast to the short hair on the chin of those species that belong to genus *Pithecia*. Bearded sakis have moderately long hair that is parted down the middle of the head. The nares are separated by a broad septum and even open somewhat upward. The relatively small ears are nearly imperceptible in the fur. Fur of the bearded sakis is flat and sleek on the body, but the shoulder region and upper arm give the impression of a broad cape because the fur hairs are long and dense there, just as are the beard and head hair. The tail gives the impression of being very thick because the tail fur is dense and long. It has a blunt tip and does not taper to its end, as in *Pithecia*. The body size varies between 40–46 cm, the length of tails approximates 35 cm. *Chiropotes* move in a quadrupedal manner and their hindlimbs are longer than their forelimbs, with an intermembral index about the same as *Pithecia*. Because bearded sakis are even rarer than *Pithecia*, there is little information about group size, territoriality, breeding, or social structure. The sakis are active during the day, and their food is said to consist primarily of fruits and seeds.

The third genus of Pithecinae, *Cacajao*, is commonly subdivided into three species: *C. melanocephalus*, *C. calvus*, and *C. rubicundus*. The common name of these monkeys is "uakari," pronounced wa-ke-ree. "Uakaris" only occur in tropical rain forest within a small area of the Amazon basin. They are said never to come to the ground. In contrast to the sakis, the fur of uakaris is long and thin. Parts of the body, such as the chest, are nearly naked; the face is bare, and the forehead completely bald. In two species these bare areas are bright raspberry red and give these uakaris perhaps the strangest appearance of any New World primate. Only one species, *C. melanocephalus*, has a dark face and a hairy forehead. The two other species, *C. rubicundus* and *calvus*, have red faces. The snout is blunt, the internasal septum very broad. The narrow slit of the mouth with somewhat downward projecting corners gives these monkeys a seemingly glum appearance. The body length of uakaris varies between 36–48 cm, relative to body length the tail is short, 15–18 cm long. These primates move in a quadrupedal manner; their intermembral index is the same as in *Pithecia*. Uakaris seem to form large groups. Little is known about such matters as breeding seasons and behavior, territoriality, and so on among uakaris. Their activity period is diurnal. The diet of uakaris combines fruit with leaves, buds, and seeds, and may be devoid of animal protein.

As already stated, most authors separate the howler or howling monkey as a distinct and primitive subfamily Alouattinae, with only one genus *Alouatta*. This genus covers five species, *A. villosa, seniculus, fusca, caraya,* and *belzebul*. Species of this genus are widely distributed over South and Central America east of the Andes, south to the Gran Chaco area, and north into the coastal forest of Mexico with the exception of Yucatan. The faces of howler monkeys are not furry, but their head hair extends far down on the forehead; male howlers have beards. The head of howlers appears

Figure 2-16
Alouatta palliata palliata (*A. villosa*) mother carrying child. [Photo courtesy of Ken E. Glander.]

somewhat elongated. This "elongation" is caused by the enlargement of the laryngeal apparatus and the concomitant change of the entire morphology of skull and mandible. In addition, the upper part of the breast bone is bifurcated to accommodate the enlargement of the vocal apparatus. In male howling monkeys, the size of the hyoid box is much greater than in females. This box serves as a remarkable resonating chamber that amplifies the male territorial calls that in turn gives these monkeys their names. The internasal septum is comparatively narrow in howlers, and the nasal openings are directed frontally and slightly upward. Fur color dimorphism occurs in *A. caraya:* males are black, and females and young of both sexes have olive colored fur. Howler monkeys are among the largest New World monkeys. Their body size varies from 40–70 cm, the long tails measure between 50–75 cm. Howler monkeys have truly prehensile tails; the last third of the tail has a well-developed friction pad on the ventral aspect. Fore- and hindlimbs are nearly of equal length, intermembral indices range from 92 to 105. Howler monkeys are mainly arboreal animals. They feed in the peripheral branches of trees and often jump from outer branches of one tree down into those of adjacent trees, having all "five" extremities extended to grasp the small branches below when landing. *A. villosa* have been seen to "bridge" gaps between larger branches with their bodies to make passage for others. Howlers have occasionally been observed swimming. In general, they are not very elegant in their movements but move in a deliberate climbing way with all "five" limbs, often walking quadrupedally on thicker branches. Like other cebids, they do not grasp between thumb and index finger but between index and third finger, a pattern that can frequently be observed in more generalized animals such as the opossum (Erikson, personal communication). These diurnal animals are very specialized in their diet: they are leaf-eaters like the Old World colobines but lack the sacculated stomach of the latter group. Howlers also eat fruit, flowers, birds, rotten wood, and bark, in addition to leaves. Howlers have often been studied in the wild. The first modern field study of a primate was, in fact, an extensive account of their behavior by Carpenter (1934). These monkeys usually live in moderate-size groups of ten to twenty individuals. In addition to defining and defending territories, the howling may also help to locate different groups and possibly has other social implications (group coherence, for example). Howlers may also shake and tear branches when excited. Hierarchical dominance patterns are present in both adult males and adult females, with the younger adult females dominant, at least, among females (see Glander 1980). Howlers are born throughout all seasons of the year. Single young are the rule but twinning has been observed.

Saimiri—the squirrel monkey—is commonly regarded as having two species: *S. sciureus* and *S. oerstedii*. Squirrel monkeys are widely found in tropical rain forests of South America east of the Andes. These monkeys seem, however, to prefer living in gallery forests or at forest edges. In contrast to the other small-bodied South American monkeys, species of *Saimiri* have more elongated, egg-shaped heads. Squirrel monkeys have a white masklike area around the eyes, but the snout is usually dark. The ears are comparatively large and often tufted. A wide internasal septum separates the nasal openings on the short and blunt snout. The tail has a tuft of long hairs on the tip, whereas the fur is dense and short over the tail length, as it is also on the body. These comparatively small and nervous primates measure about 20–40 cm in body length, the tail longer than the body, varying between 35–47 cm. *Saimiri* species are the smallest of the Cebidae. The hindlimbs are considerably longer than the forelimbs, with an intermembral index averaging about 77. These small cebids are very agile and run and leap quadrupedally. Evidently the breeding season is restricted. *Saimiri* is reported to congregate in very large numbers (up to 500). They are diurnal and their diet consists mainly of fruit and insects. Squirrel monkeys are easily kept in captivity and were frequently sold as pets in the past. At present, numerous *Saimiri* colonies are maintained by research laboratories.

The capuchin monkey—genus *Cebus*—gives its name to the entire superfamily, Ceboidea. There are two sorts of capuchins that differ most noticeably in the arrangement and length of hair on the head. One group constitutes the tufted capuchins, whose elongate hair on the top and both sides of the head forms paired tufts of individually variable size and shape. Tufted capuchins all belong to the species *C. apella*. The three remaining species, *C. albifrons, C. capucinus* and *C. nigrivittatus* have dark head hair caps without tufts. This latter trio of species includes slightly smaller sized animals than does *Cebus apella*. Primates of the genus *Cebus* are found in most of the forest areas of Central and South America; they live in mountainous regions up to an altitude of about 7,000 feet. Fruit and insects are combined in the diet of these diurnal primates. The body size ranges from 30–60 cm, the tail length varies also between 30–60 cm. The tail is prehensile but is not equipped with a friction pad at its tip. Basically, capuchin monkeys are quadrupedal climbers. In this locomotion the tail is always involved as an additional grasping limb securing their movements. Capuchin monkeys live in groups, reportedly of about ten to thirty or more animals. No defined territoriality rules the life of *Cebus* groups. Capuchin monkeys have a wide variety of vocalizations. They are said to have a complicated dominance ranking order. Possibly the unusual variety in vocalization relates to the latter. The sexes cannot easily be recognized in the field, for females have a pendulous clitoris that at a distance resembles the penis. The females give birth during all seasons. Capuchin monkeys can easily be kept in zoos and are known for remarkable manipulative skills and good memories.

The subfamily Atelinae includes three genera: *Ateles*, the spider monkey; *Brachyteles*, the woolly spider monkey; and *Lagothrix*, the woolly monkey. The genus which gives its name to this subfamily, *Ateles*, is subdivided into four species: *A. paniscus, A. belzebuth, A. fusciceps*, and *A. geoffroyi*. Spider monkeys occur in tropical South and Central America and range north as far as Mexico. The head of spider monkeys is rounded and the forehead is high, whereas the snout protrudes somewhat but is not pointed. A wide internasal septum separates the nasal openings and the face between the eyes. The nose and mouth are usually naked. The ears are large and positioned relatively low on the head, and hidden entirely in the shaggy head hair in some species. The fur is variable in texture and color both within and between different species. Species of *Ateles* are the largest of the South American monkeys, with body length ranging between 35–65 cm, and tail length between 60–90 cm. Long, slender limbs and a relatively short trunk with rounded belly add to the spiderlike appearance of these monkeys and gave rise to their common name. Spider monkeys live in the high canopy of rain forests and typically move with all five extremities, for they have a very long prehensile tail bearing a large friction pad on the ventrodistal end. Spider monkeys are highly adapted for arboreal life, and they use all four limbs for climbing, at least two of the limbs holding on most of the time. Sometimes they do, however, hang from their tails alone. Spider monkeys swing by their arms comparatively often when on the move. Like the apes, their forelimbs are longer than hindlimbs (intermembral index averages 105); in addition both limb-pairs are long if compared to trunk length. Frequently when in captivity, they walk on the ground bipedally. The thumb is greatly or even totally reduced. Mainly fruit and nuts plus leaves and flower buds combine to make their vegetarian diet. Hierarchies of dominance are not clearly defined.

Brachyteles—the woolly spider monkey, known from only one species—is not regarded as a distinct genus by all primatologists, for these monkeys very much resemble the genus *Ateles*. The differences that have been listed are the woolly character of the fur, the comparatively narrow internasal septum, and isolated occurrence in a very restricted forest area of southeast Brazil. Perhaps the most important distinctions are a number of dental differences. Diet, activity periods, and most of the known skeletal and nondental anatomical characteristics are like those of *Ateles*. Also, the number of chromosomes in *Brachyteles* has been reported to be the same as in *Ateles*. The group size appears to be influenced by the availability of food and varies from family groups up to assemblages of 100 individuals or more. No breeding season seems to occur, because spider monkeys have been seen carrying the young during all months. Twinning is rare. Females and males cannot be easily discerned because the clitoris is long and pendulous.

Woolly monkeys—genus *Lagothrix*—are given their common name because of their dense, soft, woolly fur. Most primatologists accept two species, *L. lagothricha* and *L. flavicauda*. The latter species occurs only in a very restricted area in western Peru. *L. lagothricha* are found in central Brazil northward into southern Columbia and Venezuela. Woolly monkeys live in rain forests, ranging into mountainous regions up to about 3,000 meters. The body length of woolly monkeys varies between 40–60 cm., the powerful prehensile tail between 55–75 cm. Woolly monkeys are very skillful in the use of their prehensile tail, which has a friction pad along its ventrodistal third. Their limbs are nearly equal in length, the forelimbs somewhat shorter than the hindlimbs (intermembral index averages 98); and like the other monkeys with prehensile tails, they locomote with all "five" extremities. Woolly monkeys are arboreal and when climbing, the tail almost always keeps contact with branches. When on the ground they frequently adopt a bipedal posture, and when standing, the tail is used as a tripod strut. The diet of these diurnal animals combines mainly fruit, nuts, leaves, flowers, buds, and insects but will include also small birds and mammals when available. Woolly monkeys live in groups of moderate size (about 25) and often join species of other cebid genera and forage together. Almost nothing is known about a possible breeding season, territoriality, or the social behavior of members of this genus.

The second family of New World monkeys includes only one genus and species. This species, *Callimico goeldii* or Goeldi's marmoset, is another rarely observed platyrrhine. *C. goeldii* lives

Figure 2–17 *Lagothrix lagothricha.* [Photo courtesy of Christian R. Schmidt.]

only in the remote rainforests near the tributaries of the Amazon river in northwestern Brazil, southeastern Peru, and the Patumayo of Columbia. The fur of these small monkeys is dense and silky, with hairs rather long. The color is dark brown, nearly black, with light or golden-brown hairtips. On the head the long hair forms a hairdo that resembles a dark wig, long and straight on the back and sides, and short and brushlike on top. Callimicos weigh at an average of 450–500 grams, and their bodies are about 20 cm long. Their tails measure 25–27 cm in length and appear to be rather thin, covered with long straight hairs. The intermembral index ranges from about 70–80. Goeldi's marmoset walk, run, and leap or spring sometimes in vertical clinging and leaping fashion. Callimicos have claws on all digits with the exception of the hallux. Social groups are small, consisting of a mated pair and their offspring. Goeldi's marmosets usually have single births. They have been observed eating fruits, berries, and insects.

The third family of New World monkeys, the Callitrichidae (also called Hapalidae by some authors), includes the clawed and small sized New World primates: the marmosets and tamarins. These primates are unique by having claws on all of their digits except the hallux and in having a dental formula of

$$\frac{2132}{2132}.$$

Hershkovits (1977) attributes four genera to this family. Proposed generic names, however, vary in an amazing way. It seems the Callitrichidae have provided a field day for taxonomists. However, it is not advisable that scientific names in common use among scientists continually change nor should groups be split up into too many unnecessary genera. To give an example of the confusing number of names that have been applied to two of the main kinds of marmosets, some of the proposed generic names for the genera *Callithrix* and *Saguinus* are:

Saguinus:	*Callithrix:*
Leontocebus	*Arctopithecus*
Callithrix	*Hapale*
Cercopithecus	*Jacchus*
Hapale	*Liocephalus*
Hapanella	*Mico*
Jacchus	*Miocoella*

Saguinus:	*Callithrix:*
Leontideus	*Midas*
Leontopithecus	*Ouistitis*
Marikina	*Sagouin*
Midas	*Simia*
Mystax	*Sylvanus*
Oedipomidas	
Oedipus	
Saguinus	
Seniocebus	
Simias	
Tamarin	
Tamarinus	

Many current data about the members of Callitrichidae can be derived from Hershkovitz (1977).

The genus *Callithrix*—the marmosets—is usually subdivided into three species. Marmosets occur in equatorial rain forests south of the Amazon and range south into the Mato Grosso region. With marmosets, colors of the dense shiny fur are variable as is the hair length, and, for example, many of the marmosets have hair tufts hiding their ears. The head is globular, the snout short and blunt, the forehead low and flat, and marmosets have a typical platyrrhine internasal septum. The body size of these primates varies between 15–25 cm, tail length between 25–40 cm. As with the other callitrichids, all digits except the big toe have claws. However, these "claws" are regarded as being modified from nails and thus are not true claws. Only the big toe is equipped with a fully flattened nail. The hindlimbs are longer than the forelimbs (Figure 2–19). The intermembral index of marmosets averages 76, almost the same as modern humans. Marmosets live in small social groups consisting of a mated pair and their offspring. Female marmosets predominantly give birth to two offspring, and the male takes a prominent role in carrying and protecting the infants. These agile, diurnal primates have a variable diet consisting primarily of insects, fruit, gum, and sap. *Callithrix* have been seen to gouge bark off trees with their incisors and canines in order to get gum and sap.

The tamarins—genus *Saguinus*—are the most common callitrichids, and the group includes at least twelve species. They inhabit most of the Amazon basin, the coasts of Columbia, and they range into southern Central America. The pelage of this genus is quite variable, varying from white body fur with gray face in *S. fuscicollis*

Definition of Order Primates

Figure 2–18 *Cebuella pygmaea,* pair with child. [Photo courtesy of Dieter Glaser.]

to black with white moustache in *S. nigricollis*. The head is long, and ovoid with a short blunt snout. The body size of tamarins ranges from 15–31 cm and the tail from 23–43 cm. The hindlimbs are longer than the forelimbs and the locomotor behavior of tamarins is primarily quadrupedal walking, running, and bounding, although leaping is also common (Fleagle and Mittermeir, 1980). Tamarins live in dense forests and spend a predominant amount of time in the lower and middle levels. They often sleep in holes in trees. They live in small social groups ranging from two to twelve, consisting of a mated pair and their offspring. The usual litter size is two offspring. Their diets include plant material, gum, sap, fruit, and insects. The vocalizations of tamarins are high-pitched, birdlike notes.

Cebuella pygmaea—the pygmy marmoset—is the smallest of the South American monkeys and, in fact, the smallest of all monkeys. Pygmy marmosets live in the dense tropical rain forest near the Amazon river. The fur of pygmy marmosets is yellowish brown and whitish on the ventral surface. The overall color impression is grizzled gray, often with a greenish tint. The tail is faintly cross striped. Comparatively large ears are hidden in the fur. The body of this small species is about 13 cm long, and it has a tail that measures close to 20 cm in length. These animals can weigh between 120–190 grams when grown up, just "one handful" of monkey. Together with *Galago demidovii* and *Microcebus murinus*, these species also hold the record of being the smallest primates. Pygmy marmosets live in family groups consisting of an adult couple together with offspring of two generations (or more). Pygmy marmosets sleep in tree holes; usually they have twin births. As with many callitrichids, infants ride on their parents and on older siblings and often only change over to the mother to nurse during the first three months of life (Christen, 1968). The pygmy marmoset eats gum, sap, insects, arthropods, and fruit. Pygmy marmosets have fairly procumbent incisors and canines, and it has been suggested that this is an adaptation for gouging off bark to get at tree sap and gum. It also appears that the distribution of the pygmy marmoset is limited by the distribution of the trees producing the sap and gum they eat. The primary locomotions used by these primates are leaping and quadrupedal running like squirrels. The vocalization of pygmy marmosets is a chirp that resembles that of a cricket (Figure 2–18).

The genus *Leontopithecus*—the lion tamarin—includes only one species, *L. rosalia*. Lion tamarins occur mainly in the tropical rain forests of eastern Brazil. These monkeys have dark, hairless faces with long manes that conceal their ears. Their fur varies in

Figure 2–19 *Callithrix jacchus*. [Photo courtesy of Michael D. Stuart.]

color from golden and reddish orange to black. The lion tamarins are the largest living callitrichids with a body length ranging from 22–37 cm and tail length ranging from 30–36 cm. The intermembral index averages over 80, a little higher than in other callitrichids. The hands and feet of tamarins are narrow with unusually short thumbs. Lion tamarins are pairbonded, and they often give birth to twins. Like many of the other callitrichids, the male carries the offspring. They sleep in hollow trees on thick tangles of epiphytes. Their diet includes fruit, insects, and animal matter. Their vocalizations include shrill whistles and chirping. Because of the golden hair and lionlike appearance, members of this species were once exported as pets in large numbers, but only a small fraction survived removal from the wild. Today, the lion tamarin faces imminent extinction in the wild.

For those who want to read more about New World monkeys, much of the more recent literature is listed in Moynihan (1976) and Hershkovitz (1977). Earlier coverage is given by Napier and Napier (1967) and in three volumes by Hill: Primates III, IV, and V.* Nevertheless, a concise, up-to-date summary of the platyrrhines, similar to several available for other major groups of primates, has not been produced.

Cercopithecoidea

All Old World monkeys are included in the superfamily Cercopithecoidea with one family, Cercopithecidae, that has in turn been subdivided into two subfamilies, the Cercopithecinae and the Colobinae. The colobines are mostly arboreal, eat leaves and have sacculated stomachs. The other Old World monkeys with subequal fore- and hindlimbs, simple stomachs, and cheek pouches are assigned to the subfamily Cercopithecinae.

Species of the most common genus of the Cercopithecinae, genus *Macaca*, called "macaques", also have the widest distribution. In Africa, macaques live in Algeria and Morocco. The predominantly African species *M. sylvanus* is the only extant primate inhabiting what is technically part of Europe, the Rock of Gilbraltar. Even so, some believe the "barbary ape" may have been introduced to Gilbraltar by man.

Most primatologists refer twelve species to the genus *Macaca*. Eleven of these species range across eastern Asia, from India,

*Hill's information appears to be rather arbitrary at times.

Ceylon, Afghanistan, and Tibet to China and to many islands including Java, Sumatra, Borneo, Celebes, the Philippines, Formosa, and the Japanese Islands. Artificial colonies of macaques exist in the Western Hemisphere. An example is the colony of *M. mulatta* on Cayo Santiago Island off the coast of Puerto Rico that has been under constant scientific observation for many years. Macaques are adapted to numerous environments, and some even inhabit areas with occasional snowfall and frost. Others live on seashores. Macaques thrive in tropical rain forests; they hide in rocky mountainous areas and invade villages, cities, and temple districts where they adjust easily to more "civilized" environments. Macaque body lengths range between 34–70 cm, the tails of different species showing different degrees of reduction from very short stumps (*M. maurus, M. fuscata, M. speciosa, M. sylvanus*) to middle-sized and long-tailed forms (*M. sinica, M. radiata, M. fascicularis*). Tail length differs also widely within and between species; see Fooden (1971) and Wilson (1972). Limbs are close to each other in length with forelimb averaging about 90 percent of hindlimb length; intermembral indices vary from about 85–95.

Those macaques that live in cold environments have long and dense furs. There is considerable sexual dimorphism in body size and, partly correlated with this, in the size of canines: male macaques are bigger than females and have longer canines than the latter. Fur color and arrangement vary considerably with the many species, as does the degree and color of sexual swellings of female macaques in estrus. Those external changes of females do not occur in *M. radiata* and *M. sinica*. In male macaques a higher intensity of coloration of sexual skin also develops during periods of sexual activity. Macaques have protruding, blunt muzzles. The internasal septum is often very narrow and the nasal openings project downward and somewhat laterally. The eyes are typically positioned comparatively near to each other. Cheek pouches enlarge the cheek region and protrude considerably when filled with food. Differing from the females, male macaques have larger tori (brow ridges) above their eyes. The ears of most macaques are pointed at the upper end, forming the so-called "Darwinian" angle. The face of macaques is nearly hairless, but some have patterned arrangements of hair on top of the head. In all macaques (except *M. maurus*) the ischial callosities are separate in the midline. Macaques are fully capable of grasping, having an opposable thumb (as do all Old World monkeys). They frequently engage in mutual grooming, especially in connection with sexual activities but also in mother-child relationships. Relatively short and robust limbs of nearly equal length characterize these monkeys, for the intermembral index averages close to 90. Macaques are quadrupeds and are very able climbers. One species, M. *silenus*—the wanderoo—prefers arboreal habitats. The amount of time spent in trees and on the ground varies intra- and interspecifically. Macaques occasionally exhibit a bipedal stance, especially when peering at far-away objects. They can walk or run over some distance bipedally across the ground, as do many primates that are partly terrestrial. Numerous papers have been published on macaque behavior. Macaque group sizes are highly variable according to the species concerned, differences in the environment, and other factors. No group contains fewer than two adult males. Group size normally varies from ten up to around forty individuals of different ages. Relations of dominance are clear-cut between adult males in a number of macaque species. Macaque groups evidently do not live in distinctly circumscribed territories that are defended, and in addition, different adjacent troops appear to avoid fighting. An interesting exception are groups that live in urban environments. These engage in serious and frequent battles (see Southwick et al., 1965). Sexual behavior, pair-bond relationships, and times of high reproductive activity differ widely between the different species of macaques. Learning plays an important part in their ability to adapt easily. Thus, two species include a large amount of crustaceans and mollusks in their diet: *M. fascicularis* and *M. cyclopis* swim and dive to obtain their seafood. Diets throughout the many members of this genus are quite variable and include fruit, nuts, grass, leaves, and birds as well as the shellfish previously mentioned.

Cynopithecus niger—although a cercopithecine monkey—has the common name "Celebes black ape." The genus is monotypic. Black apes are found only in the northern areas of Celebes, on some adjacent small islands, and also on one of the Moluccas (Batjan), where it presumably has been introduced by man. As the common name expresses, the fur and body of these monkeys is dark, nearly black. The limbs are subequal in length (intermembral index over 92). The tail is reduced to a short stump. Celebes black apes have a body length that ranges between 50–80 cm. Looking more like that of baboons than that of macaques, the snout is elongated and has anteroposterior deep grooves on both sides. A prominent brow ridge above the eye region also resembles baboons rather than macaques. Like the body, the face is black; the internasal septum is narrow and the nasal openings slitlike. A tuft of long and straight, upward and backward directed hairs on top of the head typifies this monkey. Little is known about the behavior of *Cynopithecus*. These mon-

Figure 2–20 *Cynopithecus niger.* [Photo courtesy of Dieter Glaser.]

keys live in tropical rain forests, are diurnal and feed mainly on fruit.

Seven species of genus *Papio* populate most of sub-Saharan Africa except some small areas in central West Africa (Ivory Coast). The north-western species *P. hamadryas*—the sacred baboon—has a range extending across the Aden channel of the Red Sea into Arabia and thus into the edge of the Asian continent. Two of the species inhabiting rain forest areas of West Africa (Cameroon, Congo, Spanish Guinea), *P. sphinx* and *P. leucophaeus,* are excluded from genus *Papio* and regarded as a separate genus (*Mandrillus*) by some scholars. These two species, commonly called mandrill (*P. sphinx*) and drill (*P. leucophaeus*), live in the rain forest but spend much of their active daytime on the ground. Mandrills and drills have swollen areas running along each side of the upper part of the snout. Each swelling has four or five bright blue parallel ridges in adult male mandrills. In contrast, the entire nose is bright red. This facial coloration of mandrills is perhaps the most striking of any

Definition of Order Primates **55**

mammal. White tufts of hair in the cheek region and a yellow beard add to the amazing impression. In both mandrills and drills, the face is bare around the eyes and on top of the muzzle. Both species have comparatively prominent and large brow ridges. The ears of mandrills are light colored and protrude slightly from the fur, whereas they are hardly visible in drills and are colored black. In contrast to mandrills, drills have plain black faces, white beards, and white cheek fur. Contrasting red and blue colors on the perineal region of male drills are presumably comparable, as sexual signals, to the facial and perineal colors of mandrills. Male mandrills have brightly colored genital regions, showing red, pink, blue, scarlet, and purple hues. Female mandrills and drills do not have these striking skin colorations. The fur of mandrills is thick, coarse, and dark grey. Drills have greenish-grey furs, and in both species, the ischial callosities are pink. Females are smaller than males, and the body length ranges from around 65–80 cm, tail length between 7–12 cm. Females show moderate sexual swellings that are not especially colorful. Forelimbs and hindlimbs are close to equal in length, but the forelimbs seem longer and more muscled. Because these two species are much more arboreal, their digits are also longer than those of typical baboons. They accomplish a functional elongation of their forearms by walking with the hands in digitigrade position. Although they may be of the same genus with the baboons and are certainly closely related, drill and mandrill are not typically spoken of as baboons. Drills and mandrills are omnivorous, with fruit as the predominant food.

Baboon species—genus *Papio*, in a strict sense—have adapted to various habitats ranging from tropical forest to semi-arid savannahs. The deep and long snout of baboons led to coining the collective noun "cynomorpha"—"doglike animals"—for baboons. Unlike drills and mandrills, the long snout of baboons is comparatively narrow in the nasal region, and the wide nasal openings point forward. In *P. hamadryas* the tip of the nose is slightly upward tilted, or snub-nosed; their face is naked, as are the ears, which, however, are hidden under long hairs in males. The dense fur varies through greyish, green, yellow, or brownish tones between the various species of *Papio*, and males often have very long hair around the shoulders, like a cape that has the effect of seeming to enlarge this area. This fur cape is especially well-developed in *P. hamadryas*. Hairs of the cheek region are also often long in male baboons. The body length varies from around 50–80 cm, tail length between 50–60 cm. Sexual dimorphism in body size of *Papio* species is marked; the males are often almost twice as large as their females

(Figure 2–21). Females have pronounced sexual swellings of the perineal region when in estrus. Fore- and hindlimbs are approximately equal in length with intermembral index ranging up to 100. The tail is of variable length in different species but is comparatively longer than in mandrills or drills. Baboons are terrestrial and regularly walk quadrupedally, placing the hands in a digitigrade position; they frequently adopt a bipedal stance, especially when "lookouts" stare across grasslands in search of danger. Baboons often sit on their large ischial callosities that, in *P. papio*, are bright red. Numerous publications deal with the social behavior of baboons, and these animals are consequently among the best-documented series of primate species. Social behavior varies widely within and between species, as does group size and composition of social groups. Many of these behavioral differences seem to be

Figure 2–21 *Papio cynocephalus.* [Friderun Ankel-Simons photos.]

correlated with differences in the environment. All baboons live in large, mixed groups except hamadryas baboons, which gather in one-male "harem" groups. Adult female baboons almost always outnumber males in the troop. Even so, the males are the leaders, with numerous social functions. Dominance is an important factor for the baboon of both sexes. Mutual grooming is frequent and makes up a significant fraction of all social activity. Newborn baboons occur throughout the year. Births are, however, especially numerous during October through December in South and East Africa. Baboons usually retire into trees or onto steep and rocky cliffs at dusk and sleep in large colonies; thus, individuals are relatively safe from predation. At dawn they leave these sleeping places for the day's foraging. The diet of baboons includes fruits, nuts, berries, seeds, roots, and small animals.

The genus *Theropithecus* with only one species, *T. gelada*, is called the gelada or "bleeding heart" baboon. Geladas occur only in a comparatively small mountainous area in Ethiopia. This region is at high altitude, rocky, and devoid of trees or other types of dense vegetation. The snout of these somewhat baboonlike monkeys is comparatively shorter from front to back and higher than the muzzle of *Papio* species. The nasal openings are not situated quite as far forward on the snout as in *Papio* but are somewhat triangular and tilted upward. A not very prominent supraorbital torus makes the eyes appear deep-set. The fur is coarse and is predominantly dark brown to buff. Males have a long mane, and the tail tips are tufted like lions in both sexes. As in the other baboons the face is naked. The ears are positioned comparatively far back on the head, where in male geladas they are hidden in the long fur but are visible in females. Perhaps the most striking characteristic of geladas is a more or less triangular, pale pink, or red naked area on the chest, which is exaggerated in its appearance by surrounding skin knobs in female geladas (Figure 2–22). This area is the so-called "bleeding heart." The skin changes its color in females when they come into estrus. Females also have skin knobs around the perineal region, and these also enlarge at times of sexual activity. Size range in geladas overlaps with that of baboons (body length 50–75 cm). The intermembral index is in the 90s, as among *Papio* species. Tail length varies between 30–50 cm. Females are smaller than males. This dimorphism is, however, less marked in geladas than in common baboons; this is clearly one of the most terrestrially adapted of all the monkeys.

Geladas live in large groups ranging up to about 400 individuals at certain seasons. These groups are composed of numerous small

Figure 2–22
Theropithecus gelada with typical naked breast adorned with skin knobs. [Photo courtesy of Dieter Glaser.]

core units that can consist of one male with four to eight females plus offspring, of groups of exclusively subadult males, or of play groups of juveniles of both sexes. The size of the core units ranges from five to around thirty individuals. These animals sit most of the day on their well-developed ischial callosities, digging up roots and collecting small food items, even dragging themselves around quite a bit in this position. Large groups only keep together when food is abundant. During the dry season they tend to split up into the core families. Thus, group size and composition are widely influenced by environmental factors. When traveling long distances in a quadrupedal manner, females and young are kept in the center of the herd or more in a wing that is between the group and a cliff-edge and is thus protected from possible intruders (Figure 2–23). In geladas, like other baboons, the lids of the eyes are white and can be exposed by rapid backward movements of the scalp. This ability, together with other facial expressions such as exposure of the gums and

Figure 2-23 Group of *Theropithecus gelada.* [Photo courtesy of Dieter Glaser.]

teeth, is important in gelada social behavior. Newborn geladas are abundant during February, March, and April. Terrestrial feeding on grasses, roots, and other small objects has made for large cheek-teeth and small incisors, a condition reversed in arboreal forest forms like drill and mandrill.

Cercocebus—the mangabeys—includes five species. The mangabeys are found in rain forests all across tropical central Africa. Two species of these monkeys seem to be attracted by swampy areas and stay predominantly in trees (*C. albigena, C. aterrimus*). The other three species (*C. torquatus, C. atys* and *C. galeritus*) are sometimes terrestrial when not feeding and sleeping. The head and face of mangabeys appear to be shorter and rounder in outline compared to those of macaques. Hairtufts around the face, contrasting colors, and different "hairdos" are characteristic of mangabeys; they have chalk white areas above the eyes, although their face is naked or only sparsely haired, and the nasal openings are directed somewhat laterally. *C. torquatus, galeritus,* and *atys* have short fur, whereas *C. albigena* and *C. aterrimus* have long hairs. All mangabeys have well-developed ischial callosities. Mangabeys are elegant and slender monkeys of around 45–58 cm in body length, with long tails of lengths varying between 64–89 cm. The intermembral index is fairly high around 86 as in *Cercopithecus*. Males tend to be bigger than females. *C. albigena* and *C. aterrimus* have a tendency to put the hands in a digitigrade position when walking on the ground. This behavior seems to be less pronounced in the three remaining *Cercocebus* species. One species, *C. torquatus*, has even been reported to be a good swimmer (Malbrant and Maclatchy, 1949). Field observations about this species of the genus are incomplete and to some extent contradictory. Thus, there are reports both on intensive vocalization and of unusually quiet mangabey population groups. Chalmers (1968) reported on the considerable resemblances between mangabeys, baboons, and macaques in vocalizations and facial expressions. He considered the black mangabey adapted to communicate under poor visibility conditions in the forest; its calls are commoner than gestures, but the gestures also show exaggerated movements. Mangabeys move around in groups of twenty to forty individuals during the day but split up into smaller units (around five individuals) to spend the night. Female mangabeys have only slight color and tumescence changes of perineal skin. Their predominantly vegetarian diet is said to be mostly composed of fruits and nuts, and mangabeys do most of their feeding in trees.

The genus *Cercopithecus*—the vervets or guenons—is one of the most diversified genera of primates; around twenty-two valid species appear to exist, and many more than this number have been proposed. Guenons are found widely distributed throughout sub-Saharan Africa in different environments. There is a high degree of variety in fur thickness and length, coloration, and contrasting adornments such as beards, hairtufts, and bright color spots. The snout is less prominent in this genus than in the other cercopithecines already described. Usually the face is hirsute except around the eyes and on the nose. Internasal septa are comparatively broad in some species. The body length of the members of this genus varies between 30–65 cm. All guenons have long slender tails that are always longer than the entire body, tail length varying from 50–110 cm. The intermembral index typically averages in the mid-80s. Male guenons are slightly larger than females. They have small ischial callosities. Guenons mainly locomote quadrupedally. Guenons have diurnal activity rhythms, but there is one species, *C. hamlyni*, that was once reported to be nocturnal. Areas covered by daily travels of *Cercopithecus* groups vary in size, but these wanderings are mainly ruled by the distribution and availability of food sources. The ratio of males to females in these groups averages around two males and three females. Relationships of dominance are not very elaborate. Social grooming is rare compared with the other, previously described genera of cercopithecines. *Cercopithecus* females do not have sexual swellings or change of skin color during estrus. Most of the species live predominantly on fruit, although *C. ascanius* eat a good deal of leaves, and two species are reported to eat many insects as well (*C. aethiops* and *C. mitis*).

Two *Cercopithecus* species are ascribed to distinct subgenera by some authors; these are the West African *C. nigriviridis* (as *Allenopithecus*) and *C. talapoin* (as *Miopithecus*). In *C. nigriviridis* the ischial callosities are somewhat more prominent than in other *Cercopithecus* species, and the body is slightly more robust. This species lives in swampy areas, which gives rise to the common name "Allen's swamp monkey."

"Talapoin monkey" is the common name of the other subgenus of *Cercopithecus*, (*Miopithecus*), containing the sole species *C. talapoin*. Whereas the size of the larger swamp monkeys range within those of the other guenons, the talapoin monkey (also called mangrove monkey) is on average the smallest of the Old World monkeys. Talapoins have an approximate body length of about 35 cm and an approximate tail length of about 37 cm. Talapoins appear in mangrove forests, swampy areas, and also on river banks in coastal central West Africa. The head of talapoins seems more spherical than those of other *Cercopithecus* species, probably because of the relatively short snout. The diet of talapoins is

Figure 2–24 *Cercopithecus aethiops.* [Friderun Ankel-Simons photo.]

predominantly fruit. Both *C. (A.) nigriviridis* and *C.(M.) talapoin* have rather restricted ranges.

Erythrocebus—the last of the cercopithecine genera to be discussed—has but one species, *E. patas*, the patas or military monkey. Patas monkeys are adapted to ground and open country, living and widely distributed in lowland savannahs and wooded steppes south of the Sahara. They are active during the day and climb into trees in the evening, where they spend the night. The muzzle of patas monkeys is moderately long, and the eyes are positioned comparatively near to each other. Adults of both sexes have a black stripe above the eyes and a white mustache. Except for a light area around the eyes, the face is hairy. Tufts of long hair on the cheeks and above and in front of the ears give the face a broader appearance than it actually has. The body and the limbs are slender, and the fur is comparatively coarse. Male patas monkeys are considerably larger than females. Males have a body length of about 65 cm; the tail is approximately the same length as the body. The intermembral index averages about 92. During quadrupedal locomotion the hands are usually held in a digitigrade position. Although all male groups occur, most patas monkeys live in groups of about five to thirty individuals that includes but a single adult male (Gartlan, 1975). The one adult male patas functions as the guard and outpost of the group. Seemingly, females are much more numerous ranging between four to twelve adult females to one adult male. Patas monkeys frequently adopt a bipedal stance and support themselves in this position with the tail also. No sexual swellings or other visual signals of heightened sexual receptivity occur in female patas monkeys. Breeding tends to fall in the months between December and February. Newborns hang under their mothers' bellies and become semi-independently precocious after only two weeks or so. Patas monkeys eat some fruit, hard seeds, grass stems, and rhizomes.

"Guereza" is the common name for species of genus *Colobus*, from which the second subfamily of the Cercopithecoidea—Colobinae—takes its name. All Colobinae lack the cheek pouches found in Cercopithecinae and differ from the latter in having sacculated stomachs (Kuhn 1964). All guerezas are restricted to Africa, but all the other colobine genera occur only in East Asia.

The genus *Colobus* is often subdivided into three subgenera: (*Colobus*) with *C. (C.)polykomos* and *C. (C.) guereza*, (*Procolobus*) with *C. (P.) verus* and *Piliocolobus* with *C. (P.) badius* and *C. (P.) kirkii*. Also, their five species are said to contain a high number (over forty) of proposed subspecies. Guerezas are widespread throughout the rain forests of Africa south of the Sahara. *C. guereza* and *C. polykomos* have black and white fur. Their basic fur and body color is black, adorned with long white hairs that form capelike structures along the body sides and on the tail. The black face is surrounded by a rosette of white fur; the head cap is black. These combinations of black and white, however, vary within and between the two large species. The black and white fur of colobus monkeys is highly prized, and because of this these monkeys are rapidly vanishing. *Colobus verus* is much less conspicuous in fur color and length of hair, and it also lacks capes and tufts. *C. badius* and *C. kirkii* are mainly black and brown-red in color. The head of *Colobus* is more globular than in most of the cercopithecines, and the snout is not very prominent. The face is nearly hairless: the ears relatively small. In the two large species (*C. guereza* and *C. polykomos*) the nose extends somewhat over the upper lip. There is little sexual dimorphism in size. In body length the five species range between 43–69 cm, and the species *C. guereza* and *C. polykomos* are larger than the others. The tail length varies between 41–89 cm. Usually the tail is longer than the body. The hindlimbs of colobines are longer than the forelimbs, with an intermembral index averaging around 80. All colobines have the thumb reduced usually to a very short stump. Ischial callosities are comparatively small in species of this genus. Colobines are highly arboreal and normally occupy the high levels of the forest canopy, except for *C. verus*, which often comes close to the ground. Guerezas seem to live in territories that are hotly defended against intruders of the same species. Groups number about fifteen animals, but no clear cut hierarchies of dominance have been observed within these groups. There is, however, a great deal of variation in social behavior among the several species of this genus. Guerezas have frequently been observed to mix with troops of guenons. Although slight degrees of sexual swelling occur at times in females, little is known about a possible breeding season. The fur of newborn *C. guereza* and *C. polykomos* is white. Mothers of *C.* (*Procolobus*) *verus* are reported to carry very young infants in their mouths. Leaves are the primary food of guerezas but *C. badius* also takes a significant amount of fruit.

Species of *Presbytis*—the langurs or leaf monkeys—are the most widespread of colobines and live across a wide range of East Asia. (India, Pakistan, Ceylon, Thailand, Malaysia, S.W. China, Indochina, Sumatra, Java, Borneo, as well as many smaller islands). At least fourteen different species (and more than eighty subspecies) have been proposed. Langurs are adapted to many different habitats;

they occur high up in the mountains, in rain forests, in mangrove thickets, in dry areas, and can range into villages and towns.

The heads of langurs are globular in outline, the snout short and the ears bigger than is typical of guerezas. In contrast to guerezas, the nasal openings of langurs are separated widely from the upper lip, that is, the nose is very short. The faces of langurs are naked or only sparsely haired. Hair tufts on crown, cheeks, and above the eyes vary the appearance of different langur species. The fur is long and color patterns are of many kinds. Newborn langurs typically differ markedly in color from adults. White circles around the eyes and white lips characterize *P. obscurus* and *P. phayrei* (Figure 2–25). These medium to large-sized monkeys all have sacculated stomachs. Body length ranges from 40–80 cm and tail length between 50–110 cm for this genus. In all the langurs the ischial callosities are relatively small and separate. Limbs of langurs are comparatively long and slender (with intermembral index typically just below 80), and their bodies are bulky. Langurs are well adapted to an arboreal way of life. However, many species spend a considerable amount of their active time on the ground. The size of langur groups ranges from a few individuals to over 100, and the sex

Figure 2–25 *Presbytis obscurus*. [Photo courtesy of Michael D. Stuart.]

ratio within these groups also varies. Langurs inhabit territories that differ considerably in size in accordance to differences of the environment. Groups are led by dominant adult males. In some species dominance behavior can be pronounced in the males, among which hierarchies are established by fighting. Infanticide of previously born young has been reported to be important when all male bands take over a group of females and kill or drive out the males previously in control. This phenomenon has been called "usurper strategy" (Blaffer-Hrdy, 1977). The adult female status within a group changes in accordance with the change of the sexual cycle. Care for young is generally very intensive, and all adult females have a part in it. Within the wide geographic range of langurs, the presence or absence of a distinct breeding season also seems to vary widely with environmental differences (Figure 2–26). Diet among species of this genus is quite variable, but in general these monkeys seem to eat fewer leaves than do species of *Colobus*. Besides leaves, langurs also feed on fruit, flowers, and even bark but seem not to add any kind of animal protein to their diets.

Only one very rare species of the genus *Pygathrix*—*P*.

Figure 2–26 *Presbytis entellus*, young animal. [Photo courtesy of Michael D. Stuart.]

nemaeus—is known. This is the "douc langur" which occurs in Laos, Vietnam, and on the island of Hainan. Douc langurs live in tropical rain forests. Individuals of *Pygathrix nemaeus* have yellowish white faces and throats, and dark eyes are placed somewhat obliquely. Chest and shoulders show a red crescent outlined in black. Extremities are black, except for *P. n. nemaeus* where legs are red. In this species the ears are less conspicuous than in genus *Presbytis* and the snout somewhat more protruding than in the latter. The body size varies between 55–82 cm, tail length between 59–77 cm. There is no marked dimorphism of size between the sexes. Little is known of douc langurs in the wild, but they have recently become somewhat commoner in zoos in Europe and America, where they seem to do well.

Rhinopithecus contains but one species, *R. roxellanae*, occurring in western China and North Vietnam at high altitudes. One of the most striking characteristic of this large sized colobine is the forward-opening snub nose that gives it its common name, the snub-nose langur. *Rhinopithecus* ranges in length between 55–85 cm, and the tails are between 60–100 cm long. Snub-nosed langurs are rather robust when compared to langurs of the genus *Presbytis*, and their limbs are comparatively short. They live in large groups, and males are as big as baboons. Their diet is reported to consist of fruit, leaves, bamboo shoots, and birds.

The genus *Simias*—also having one species, *S. concolor*—occurs only on the Mentawai Islands off the west coast of the island of Sumatra in swampy tropical forests. These monkeys have a body size of approximately 50 cm, the tail is reduced and only about 10–20 cm long. These monkeys are also comparatively robust, and the limbs are approximately equal in length. The short tail is said to be bare except for a hairtuft at its tip. Ischial callosities are comparatively large and joined medially in males.

The genus *Nasalis*—the Bornean proboscis monkey—also has but one species, *N. larvatus*. Proboscis monkeys occur exclusively on Borneo, where they live in forests and in coastal mangrove swamps. The face is nearly bare. The most obvious characteristic of proboscis monkeys is the fleshy nose that is much larger in males than in females, and that sometimes, in fully matured males, actually droops down over the mouth. In females the nose is slightly tilted upward. In comparison with other colobines, their ears are placed near to each other, their fur comparatively long, and their ischial callosities are large. The body ranges from 54–73 cm long in proboscis monkeys and the length of the tails from around 57–75 cm. Adult males are much heavier than adult females. A webbing frequently occurs between the second and third toes. Proboscis monkeys are very good swimmers and swim frequently. They have been observed to jump from trees directly into the water. Proboscis monkeys live in groups, the size of which averages around twenty animals. This species primarily feeds on leaves, but they also eat fruit and flowers.

Hominoidea

The last superfamily we have to deal with is the one that includes all apes and man himself, the superfamily Hominoidea.

Six species of lesser apes are attributed to genus *Hylobates*. These only occur in Southeast Asian rain or deciduous forests both on the continent and on many islands. Members of this genus often range into high mountainous altitudes. Generally, the head is globular in gibbons, and the ears are not very big and are often hidden in long hair. Fur color patterns are highly variable among the various species, races, and even individuals of this genus. These color patterns change with age as well. The fur of gibbons is generally rough in texture, and in some species, hair tufts are found around the bare and typically dark faces. The body length ranges

Figure 2–27 *Hylobates lar.* [Photo courtesy of Dieter Glaser.]

between 40–65 cm. Here the second highest intermembral index known among any primate occurs, ranging between 120–140. As in all the hominoids an external tail is absent in all *Hylobates* species. Gibbons have comparatively small ischial callosities. All hylobatids—gibbons and siamangs—have very long forearms and often move by arm swinging alone. When on the ground or walking on larger tree branches, hylobatids frequently progress bipedally, balancing the upright body with their long arms. *Hylobates lar* lives in family groups that consist of an adult pair with up to four offspring; see Carpenter (1940) and Ellefson (1968). Gibbons strongly defend their family territories. They have loud, far reaching melodic vocalizations that are correlated with establishing a territory as their own. Actual fighting between adjacent families seems to be rare, and there is no significant sexual dimorphism in canines or in body size. Mutual grooming occurs frequently, and infants share food. Differences in dominance are not known among gibbons. Newborn babies are present around the year: There is no discrete breeding season. They are active during the day and live on a combined diet of fruit, leaves, flowers, buds, insects, and eggs of birds or small birds.

The Siamang—*Symphalangus syndactylus*—is closely related to *Hylobates* and is sometimes called the greater gibbon. Siamangs occur only in the Malay peninsula and on the island of Sumatra. There they range into mountainous areas as do the lesser gibbons. Siamangs are black except for the snout, which is light brown in color. The hair of the coat is long, and its shaggy appearance is almost like that of a chimpanzee. Adult male siamangs have a tuft of hair in front of the genital area. Both sexes have a large naked air sac at the throat. They are 45–65 cm in body length and generally somewhat larger and more robust than gibbons. Their locomotion is like that of *Hylobates*, but the intermembral index ranges almost up to 160 and averages at 148, about 20 points higher than *Hylobates*. The generic name (*syn*, together; *dactyl*, finger) refers to the webbing of the second and third toes that sometimes reach to the proximal interphalangeal joint. Siamangs live as mated pairs or nuclear family groups, and their loud vocalizations play an important role in their territoriality. The diet of siamangs consists basically of leaves. Two more recent studies of gibbons in the wild are by Chivers (1974) and Fleagle (1976).

Orangutans occur only on the islands of Borneo and Sumatra, and the name is from Malay (*oran*, man + *utan*, forest). The one species *Pongo pygmaeus* has two subspecies: *P. p. pygmaeus* lives only on Borneo, and a second subspecies *P. p. abelii* is restricted to the western tip of Sumatra. These large apes have pronounced snouts and deep concave facial profiles. The eyes are comparatively small and positioned near each other, and supraorbital ridges are much less developed than in African apes. The upper lip is remarkably deep so that the nasal openings typically appear to be closer to the eyes than to the upper edge of the mouth. In adult males large flanges extend out from the sides of the face in the cheek region. These greatly broaden the face and thus enhance the appearance of great size, useful during threatening behavior. Orangutan faces are usually less hairy than those of the African apes, but the Sumatran orang may sometimes show the face frosted with short white hairs. The skin of the orangutan often shows a blue-grey almost metallic tint, whereas the long hair is rather coarse and its color ranges from carrot red to reddish brown. Many orangs have beautiful red beards. Also, adult males possess large air sacs on the throat that amplify their outlandish cries. The ears of orangutans are comparatively small and positioned inconspicuously. These apes are highly dimorphic, and the body length averages 95 cm for males and 77 for females. The arms are very long, much longer than the hindlimbs; the length from shoulder to wrist averages 112 percent of the length from hip to ankle (intermembral index). Orangutans usually walk on the outside of their inward-bent hands (fist

Figure 2–28 Left, female *Pongo pygmaeus*; right, male *Pongo pygmaeus*. [Photos courtesy of Dieter Glaser.]

walking), or place the hands in a plantigrade position but with fingers turned out laterally, when moving on the ground quadrupedally. These apes are mainly tree dwellers that climb about in a careful and deliberate manner, often hanging by any and all of the four extremities. However, they sometimes do also travel on the ground for some distance, especially older males (Rodman, 1973). Orangutans build sleeping nests. These nests can be more complicated in construction than those of the African apes because they contain not only a resting platform of branches but may have overhead branches pulled together above as a shelter from rain. Orangutans have only tree nests. They almost never defecate into them or nest on the ground, as gorillas commonly do. Nests are newly built each evening. Orangutans live comparatively "unsocial" lives. Adult males frequently live alone; adult females without company other than a sole offspring. It seems, however, that one old male patrols through and defends an extended territory, wherein he breeds with several females that have their own subterritories. The daily movements of orangutans are widely influenced by the availability of food in their home ranges. These apes often meet in food trees and then seem to share access placidly, although territorial conflicts between males occasionally occur. Female orangutans do not exhibit periodical swellings of the circumgenital region. The "old man of the forest" is reported to live mainly on fruits, including many different other vegetarian food items; orangutans also live on birds' eggs.

The chimpanzee is perhaps the best known of all the apes. Two species, *Pan troglodytes*—the common chimp—and *P. paniscus*—the pygmy chimpanzee or bonobo—are usually distinguished. *P. troglodytes* occurs in at least three different varieties. These subspecies of *Pan troglodytes* occur in a series of environments in central Africa, including woodland, rain forest, savannah, and also mountainous regions up to about 3,000 meters high. Much more restricted in extent is the rain forest inhabited by pygmy chimpanzees; these appear only in one region south of the Congo or Zaire River, where it arches farthest north. Because chimpanzees cannot swim, the Zaire River strictly limits the range of the bonobo on the west and north, and the Lualaba River—a tributary of the Zaire—limits its distribution on the east. Chimpanzees have large ears that vary greatly in size, outline, and position. Also, the composition of facial features is very variable; consequently, there is high individuality in chimpanzees' faces. It is clear that each of these apes learns to recognize many different individuals. The center of the face around the eyes, the somewhat protruding snout and the lips are essentially hairless. The nasal openings are positioned relatively high above the upper lip, are separated by a narrow internasal septum and vary in outline and in relative size. There is a marked superorbital torus above the eyes, and they are positioned relatively far apart. The amount, length, and direction of head hair also differs individually. For example, both sexes of common chimpanzees tend to get bald on the top of the head relatively early in life, and pygmy chimpanzees often have longer head hair with a natural part in the middle. The skin of bonobos is usually black, and it is more often flesh colored in *P. troglodytes*. *P. troglodytes* varies, however, in the coloration of the skin. The fur is usually black or dark brown except for a supra-anal white hairtuft in very young animals and the grizzled-grey, and white of old chimpanzees. *P. troglodytes* ranges in body length between 70–95 cm, females being slightly smaller than males. Pygmy chimpanzees are generally slightly smaller than the common chimpanzee. The forelimbs of chimpanzees are longer than the hindlimbs, and intermembral index ranges from about 102–114. Common chimpanzees spend a considerable amount of their daily active time on the ground. There, they walk on the knuckles (second phalanges) of the second to fifth fingers when moving quadrupedally (Tuttle, 1975). *Pan paniscus* is generally lighter and less robust than the common chimpanzee. This species tends to walk upright more often than the other, vocalizes differently and copulates frontally. (Fore- and hindlimbs of bonobos are of almost equal length, so that the legs are relatively longer than in common chimpanzees.) Pygmy chimpanzees have only recently come under observation in the wild, and what is known in regard to their behavior has been documented by Horn (1976) and Badrian and Badrian (1977). The behavior of chimpanzees in natural environments is very well documented because numerous scientists have spent, over the years, an enormous amount of time watching them in different parts of Central Africa: Goodall (1968) in the Gombe Stream area, Reynolds and Reynolds (1965) in the Budongo Forest, and elsewhere in Tanzania, Suzuki (1969), Sugiyama (1973), among many others. From these studies it appears that the behavior of chimpanzees varies considerably with different environments. Chimpanzees form groups that range in size from one or two up to around fifty individuals; these also vary considerably in composition. There are groups that consist only of mothers with infants or adults of both sexes, of males only, and mixed groups of adults and young animals of both sexes. Relationships between chimpanzees of the same group appear to be very placid, but females and young typically defer to feeding males. Female chimpanzees

Figure 2–29 *Pan paniscus* (left hand is held in a "knuckle-walking" position). [Photo courtesy of Michael D. Stuart.]

have large swellings in the anogenital skin that become bright red at time of estrus. Females in estrus usually accept copulation with several males. In Tanzania sexual activity is highest between August and October. However, there is no evidence for a well circumscribed breeding season. Group size and movements are influenced by the availability of food in the area (Nishida, 1968). Chimpanzees have an elaborate communication system by means of facial expressions, gestures, movements of the body, and vocalizations (Reynolds and Reynolds, 1965; Goodall, 1965). Social grooming is a frequent chimpanzee activity. Chimpanzees sleep in tree nests seldom more than around 4–5 meters above the ground that are newly built every evening. The chimpanzee's diet consists mainly of fruit and plant material. They do, however, also eat small mammals, including young chimpanzees, monkeys, birds, insects, and even fish when able to obtain these food items.

The largest of the extant primates is the gorilla, whose present populations exist in the highland and lowland rain forests of central and West Africa, extending east as far as the north end of Lake

Tanganyika and down into southwest Uganda. The only species, *Gorilla gorilla*, is commonly subdivided into two races: *G.g. gorilla*—the so-called "Lowland or Western Gorilla"—and *G.g. beringei*—the "Highland or Eastern Gorilla" (Figure 2–30). In addition, a third subspecies, *G.g. manyema*, (Groves, 1967) has been proposed as the Eastern Lowland Gorilla. The skin color of gorillas is black; their body hair is usually black or dark brown, but hair on top of the head is sometimes reddish. In the face, only the area around the eyes, the nose, and lips is bare, whereas the cheeks are usually fuzzy. The eyes sit deep under pronounced supraorbital tori, and the nasal openings are nearer to the upper edge of the mouth than in chimpanzees and orangutans. Gorillas have small ears that lie back on the head rather than stand off it, as in chimpanzees. The large nasal openings vary considerably in outline and size. These openings sometimes have enlarged bulging rims and thus can give particular gorilla faces highly individual distinctions. Adult male gorillas grow to around 100 cm in body length, females to around 75 cm. Male gorillas weigh about twice as much as females, so that dimorphism probably reaches its extremes with these primates. Gorillas are mainly terrestrial, except for the young that frequently climb trees. Like chimpanzees, gorillas progress quadrupedally in

Figure 2–30 Left, *Gorilla gorilla beringei*, male; right, *Gorilla gorilla gorilla*, male. [Photo courtesy of Michael D. Stuart.]

the knuckle-walking manner. They also sleep in nests, newly built each evening. Gorilla nests are usually built on the ground or low in trees. They are frequently found soiled with excrement. As with chimpanzees, the group size of gorillas is variable. Groups ranging from five up to thirty members have been counted. Old male gorillas tend to develop a coat of grey and white hairs particularly on the back, and these old "silver backed" males normally assume group leadership. Such old males, according to Schaller (1963), lead the foraging and set the pattern of sleeping and feeding times. They usually tolerate one or a few other adult males, but some males wander alone. For young males the roles within a group seem to be defined by their age, in females by the presence of infants. Grooming is normally of offspring by the mother. Young gorillas of both sexes frequently play in a quiet placid manner (Schaller, 1963, 1965). Groups usually do not move more than about one mile per day. Meetings between groups and between adult males usually lack any open hostility. Interactions among adult gorillas consist mainly of chest beating gestures, postures, facial expressions, and vocalizations that almost never lead to physical conflict. There is no restricted breeding season in the natural habitat. Young gorillas stay with their mothers for up to three years. Female gorillas have only a slight degree of sexual swelling. Recent observations on gorillas include Fossey (1974) and Goodall (1979). The gorilla diet consists mainly of leaves, shoots, roots, bark, vines, and only a small amount of fruit. They appear to be entirely vegetarian, and herbivores of this body size must consume a considerable amount of such plant food daily in order to thrive. Consequently, gorillas spend a large part of their day time activities foraging.

Homo sapiens sapiens—our own species and the only extant species (and subspecies) of the family Hominidae—is the most successful of the living primates and has widely colonized many world environments. Even so, most human populations no longer live in a natural habitat. The enormous recent expansion of our species has created an imperative task: to keep this world in a biological balance and not to drown it in an ocean of garbage and pollution. Primatologists also should consider that one of their primary concerns is to protect all primate species, including our own, from extinction.

Modern humans constitute but one highly variable species. The history and the characteristics of this species are covered in detail by other volumes in the Macmillan series in Physical Anthropology of which this book is a part. Jolly (1972) has analyzed human and primate behavior; Simons (1972) and Pilbeam (1972) have dealt with primate and human evolution; and Lieberman (1975) has written on the evolution of language. A most significant addition to the consideration of man as a primate is Brues' (1976) contribution on the biology of modern man.

Without unduly exaggerating the importance of our own species, it is fair to say that we are an unusual mammal that has evolved a number of remarkable biological features distinguishing us from all other primates.

Physiologically our ability to sweat continuously, and thus achieve high levels of heat loss under stressful conditions of exertion, is a nearly unique feature. In turn, it may be correlated with the reduction in humans of most body hair to a down. Equally distinctive is the locomotor adaptation to habitual bipedal walking in *Homo sapiens* that is both unparalleled among other mammals and unknown as to its origin. Finally, human capacities in the areas of speech and reason set off our species distinctly from other primates including the apes. Nevertheless, even in this case the size of the gap between man and animal has been decreased by the findings of many recent behavioral studies on the chimpanzee. These studies show that our closest relative can combine gestural abilities and understanding of symbols in a manner foreshadowing human capacities.

One final expression of man's distinctiveness is as a tool maker, and it is said that Benjamin Franklin first pointed out this human distinction (see Simons, 1972). Nevertheless, Goodall's work on the chimpanzees (1968) has shown that these apes might be considered to make tools, for they break off the side branches of plant stems in order to manufacture a rod by which they "fish" for termites, sticking the twig into holes in termite nests and then licking off the adhering termites. Even so, this is *not* actually a planned tool made in the same fashion as stone tools are made to a set and regular pattern. It is still true as James Boswell observed (1786) in *A Journal of a Tour to the Hebrides:* "...No animal but man makes a thing by means of which he can make another thing." In sum, man's real distinctiveness as a tool maker is not that he makes tools only, but that we make tools that are used to make other tools.

In concluding this survey of the living primates, it is well to summarize their geographic occurrence. In general, the distribution of non-hominid primates is restricted to tropical and sub-tropical areas. This is perhaps due more to their dietary needs for a continuous supply of insects, fruits, and fresh leaves than to a general inability to tolerate cold.

Tupaias and *Ptilocercus* are found in southeast Asia. Lemurs are only found on the island of Madagascar and some of the small islands nearby. The family Lorisidae has two Asiatic and two African genera. *Nycticebus* and *Loris* live in India, Ceylon, and some adjacent areas. *Perodicticus* and *Arctocebus* are found in tropical West Africa, and the genus *Perodicticus* ranges into the western parts of Kenya and Uganda. The Galagidae are spread all over Africa south of the Sahara except the most southern tip of the continent. The genus *Tarsius*—and thus the family Tarsiidae—is restricted to many southeast Asian islands.

The Anthropoidea can be subdivided into two distinct groups: monkeys of the New World and the Old World monkeys plus apes. The two families of New World monkeys live in tropical South and Central America. The most northern occurrence extends into Mexico. Among the Old World monkeys it is the genus *Macaca* that is most widespread. Macaques live all over Asia, are found in Africa north of the Atlas Mountains and include the only recent wild living monkey in Europe: *Macaca sylvanus* on Gibraltar, the most southern portion of continental Europe. Macaques are well adapted to cold climates, and in Japan they even live in areas where it occasionally snows and freezes. The genus *Papio* is widespread in Africa south and east of the Sahara, and a small Arabian population has even invaded the most western point of Asia. *Theropithecus*—the distinctive gelada baboon—lives only in Ethiopia. The genus *Cercopithecus* is found widely in Africa south of the Sahara, and the cercopithecine genera *Cercocebus* and *Erythrocebus* are also exclusively African but more restricted than *Cercopithecus*.

Of the subfamily Colobinae, only the genus *Colobus* is found in the tropical rain forest of Africa. All the rest of the family (*Presbytis, Pygathrix, Rhinopithecus, Nasalis* and *Simias*) live in eastern Asia.

The lesser apes (*Hylobates* and *Symphalangus*) are restricted to southeast Asia. The large Asiatic ape—the orangutan—lives exclusively on the islands of Borneo and Sumatra. Species of the two genera of African apes are restricted to Central Africa. And man, of course, is of world-wide distribution.

chapter 3
Teeth

Probably the most informative discussions of primate teeth are by Remane (1923, 1960), but they are in German. In English works there are the discussions of Gregory (1922), James (1960), and the series of Osman Hill (1953, et seq.). Swindler's (1976) compilation of information on variation in the dentitions of extant primates is a valuable source for data on this topic, as well as Dahlberg (1971).

All mammals have two sets of teeth, the maxillary or upper dentition and the mandibular or lower dentition. Each tooth has a crown and one or several roots. The roots are implanted into the bones of the skull (premaxillary and mandibular or dentary). The sockets for the tooth roots are called alveoli. The alveoli are usually perfect molds of the tooth roots they hold. Thus, it is possible to get some information about the size of teeth even from specimens (for example, fossils) that have lost their teeth but have the alveoli preserved. The crown of a mammalian tooth is covered with enamel. Under the enamel is dentin, which surrounds a pulp cavity. The root (or roots) of a mammalian tooth are covered with cementum. Under the cementum is dentin that, as in the crown, surrounds a pulp cavity (Figure 3–1).

The dentition of primates is, without exceptions, heterodont (*hetero*, Greek for "different"; *dont*, Greek for "toothed"). As in all toothed animals the dentition of each side of the jaw is a mirror image of the other side, across the midsagittal plane. Thus, it is sufficient to describe only one half of a dentition to know the entire structure. The dentition of a generalized primate (or mammal)

77

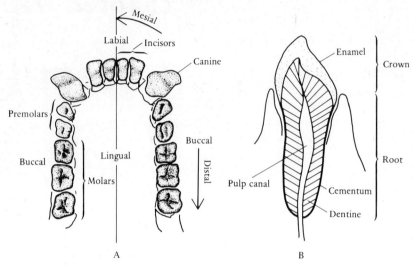

Figure 3–1 (A) Terms of orientation in teeth. (B) Basic histology, incisor.

commonly is a combination of four types of teeth. In front we find the incisors (numbering two times two in most primates, two times 3 in most generalized mammals). The incisors are followed on each side by a single canine. The canines are followed by two or three premolars (four or five in generalized mammals) on each side. Behind the premolars are three or two molars (also three in generalized mammals) that typically have the most complicated morphology of all the four types of teeth.

To clarify discussion about the dentition, it is useful to know some basic comparative terms of anatomical position. The tooth surfaces facing the cheek are the buccal surfaces, and the tooth surfaces facing the tongue are the lingual surfaces. The tooth surface oriented toward the median line in the dental arch is the mesial (or anterior) surface and the tooth surface oriented away from the median line is the distal surface (Figure 3–1). Mesial and distal also describe the positional relationship of one tooth to another. For example, incisors are mesial to canines and molars are distal to premolars.

During ontogeny some of the teeth are replaced by a second tooth generation, the so-called permanent dentition. Those teeth that are replaced are called the deciduous teeth or milk dentition. Replaced teeth are the incisors, canines, and premolars. The molars appear only in one set. Thus, there has been some discussion about whether the milk incisors, canines, and premolars make up the first tooth generation (Bolk, 1915) and the permanent incisors, canines, premolars, and molars the second generation. The alternative is that the milk incisors, milk canines, and milk premolars together with the permanent molars form one single (first) tooth generation, and only the replaced permanent incisors, canines, and premolars are regarded to represent the second tooth generation (Bennejeant, 1935). Once a tooth has been formed and has erupted, its morphology does not change thereafter except by wear or decay.

The number of each tooth type is constant for mammalian species and often for higher taxonomic groups. Because of this consistency, the number of each type of tooth for an animal can be written down as a dental formula. The dental formula is written for one side of the dentition starting at the midline—

$$\frac{\text{I-C-P-M}}{\text{I-C-P-M}}.$$

The teeth represented above the line are the maxillary teeth, and those below are the mandibular teeth. Capital letters are used for permanent teeth, whereas milk teeth are usually indicated by uncapitalized letters,

$$\frac{\text{i-c-m}}{\text{i-c-m}}.$$

The basic number of teeth in most early Tertiary mammals usually is three incisors, one canine, four premolars, and three molars. Written in the short manner of a tooth formula, this is

$$\frac{3\text{I-1C-4P-3M}}{3\text{I-1C-4P-3M}} = 44$$

and, in a juvenile,

$$\frac{3\text{i-1c-4m}}{3\text{i-1c-4m}} \quad \frac{(3\text{i-1c-4p})}{(3\text{i-1c-4p})}.$$

There is never more than one canine in each half of a dentition. Even shorter and simpler is the numerical tooth formula that can be written as follows:

$$\frac{3\text{-}1\text{-}4\text{-}3}{3\text{-}1\text{-}4\text{-}3} = 44.$$

The corresponding formula of the milk dentition is written

$$\frac{3\text{-}1\text{-}4}{3\text{-}1\text{-}4} = 32.$$

For example, the dental formula of *Homo sapiens* is

$$\frac{2\text{-}1\text{-}2\text{-}3}{2\text{-}1\text{-}2\text{-}3},$$

of *Alouatta*

$$\frac{2\text{-}1\text{-}3\text{-}3}{2\text{-}1\text{-}3\text{-}3},$$

and of *Callithrix*

$$\frac{2\text{-}1\text{-}3\text{-}2}{2\text{-}1\text{-}3\text{-}2}.$$

Milk or deciduous teeth are not always replaced by a permanent tooth. For example, the Indriidae have a small lower canine in their deciduous dentition that does not have a replacement by a permanent canine (Leche, 1896; Remane, 1960; Spreng, 1938): milk dentition

$$= \frac{2\text{-}1\text{-}2\text{-}3}{2\text{-}1\text{-}2\text{-}3}$$

and permanent dentition

$$= \frac{2\text{-}1\text{-}2\text{-}3}{2\text{-}0\text{-}2\text{-}3}.$$

However some authors believe that Indriidae have only one lower incisor and a lower canine, James (1960). If the latter proves to be true, the dental formula of the Indriidae would be

$$\frac{2\text{-}1\text{-}2\text{-}3}{1\text{-}1\text{-}2\text{-}3}.$$

Whether the Indriidae in fact do have a lower canine and only one incisor or have two incisors and no canine in the permanent dentition can only be proven with a series of specimens with milk dentition and also the fossil record of the teeth involved (Remane, 1956).

Each tooth is also given a number. An animal with the dental formula

$$\frac{3\text{-}1\text{-}4\text{-}3}{3\text{-}1\text{-}4\text{-}3}$$

has the following teeth:

$$\frac{I^1 I^2 I^3 C \ P^1 P^2 P^3 P^4 M^1 M^2 M^3}{I_1 I_2 I_3 C \ P_1 P_2 P_3 P_4 M_1 M_2 M_3}.$$

Callithrix with a dental formula of

$$\frac{2\text{-}1\text{-}3\text{-}2}{2\text{-}1\text{-}3\text{-}2}$$

has the following teeth:

$$\frac{I^1 I^2 C \ P^2 P^3 P^4 M^1 M^2}{I_1 I_2 C \ P_2 P_3 P_4 M_1 M_2}.$$

This example demonstrates some important trends in dental reduction. When reduction occurs in the incisors, usually the lateral incisors are lost first. Premolar reduction is usually characterized by the loss of the mesial (anterior) premolars. Finally, molar reduction is usually characterized by the loss of the distal molars.

Tooth morphology also shows differences of high diagnostic value, as, for example, between Cercopithecoidea (Old World monkeys) and Hominoidea (apes and men) but is of lesser use in New World monkeys. The number of teeth, for example, enables one to distinguish with a quick glance between dentitions from Old World monkeys and those from the New World. This also makes for easy recognition and separation of the members of the two different monkey families of the New World (Cebidae and Callitrichidae).

A so-called "alveolar diastema" or "apparent diastema" is found frequently in primates. This means that the sockets of two adjoining teeth are separated by a stretch of intervening alveolar bone, but that the crowns of the same teeth touch each other (e.g. orangutans' incisors). This diastema is an expression of the fact that either the crowns of the teeth are much broader than the roots or that the teeth are implanted in such a way that the long axes of two adjacent teeth are tilted toward each other. According to Remane (1960) a "real diastema" occurs when both the crowns and the rootborders are spaced apart and do not contact. This distinction is important, for one kind of diastema can be mistaken for the other in fossils where tooth crowns have been broken away.

The crowns of teeth function in different ways and according to their morphology. The incisors of extant primates are usually quite simple and conical or wedge-shaped. Canines are usually daggerlike, sharp, and pointed teeth. Premolars and molars, however, usually show large functional (or occlusal) surfaces that have differently structured and differently functioning reliefs (cusps and ridges). In very old individuals the tooth crowns can be worn down so that no surface relief remains.

Dental Morphology

As mentioned earlier, the incisors of extant primates are usually quite simple and conical or wedge-shaped. Incisors are always single rooted. Many mammals have three incisors (four sets of three); however, no primate has more than two incisors.

All upper incisors are implanted in the premaxillary bones. The premaxilla is also known as the *os incisivum*. The suture between premaxillary (also called intermaxillary) and maxillary bones is still visible in young primates. This suture makes it easy to decide how many incisors are present in the upper dentition. The number of incisors is obscured in many cases in the lower dentition because canines can be incisiform.

Members of the Prosimii, with the exception of *Tarsius*, have extremely specialized incisors. The lower incisors are tilted forward (they are then called procumbent) and flattened laterally, forming a toothcomb (Figure 3-2). The canine is frequently included in this toothcomb, and its morphology is assimilated to the shape of an incisor. The first premolar usually resembles the canine (caniniform premolar) and takes over the canine's function. This fact can easily be observed because when jaws are closed, the upper (real) canine rides down anterior to the canine shaped lower tooth and not behind

Teeth

Figure 3–2 Procumbent toothcomb.

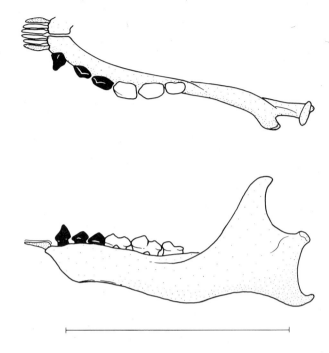

it, as it is typically the case with caniniform lower canines (Figure 3–3). In those prosimians that have a toothcomb in the front of the lower jaw, the contact between the upper and lower incisors is frequently lost. In consequence, there is often a reduction of the upper incisors (Remane, 1960). As the size of the upper incisors decreases, the size of the premaxilla also undergoes reduction. In all cases where upper incisors are lost, however, small premaxillae do remain but are reduced relative to those of a typical prosimian (e.g. *Propithecus*, the sifaka) with rather big incisors. Total reduction of the upper incisors is found in *Lepilemur*, the "sportive lemur." In

Figure 3–3 Dentition of *Phaner furcifer*.

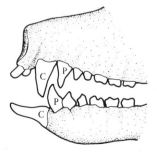

Figure 3–4 *Lepilemur leucopus*, front view of skull showing premaxilla reduced to a band around nostrils (premaxilla black).

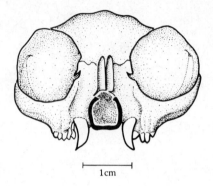

this genus the premaxilla struts across the front of the upper jaw and also embraces the nasal opening from both sides (Figure 3–4). In front of the palatine the premaxilla forms the surroundings of the large incisive foramen. Thus, the premaxilla of *Lepilemur* is the smallest among all living primates, only a clasp like, narrow bony structure (Figure 3–5). Some of the callitrichids also have procumbent lower incisors. Their incisors, however, are not as procumbent as those of most prosimians and do not form a tooth comb with the lower canines.

Canines follow distally behind the incisors. These teeth are called "canines" because of their dagger like shape—"canine" meaning the dog tooth (*canis* is Latin for "dog"). Generally, in primates the canines are larger than the incisors or premolars; canines are pointed and curved distally. In some primate genera the canines are dimorphic in the two sexes, big in males and smaller in females (e.g. *Papio*, the baboons, *Gorilla*, the gorillas); in others

Figure 3–5 *Lepilemur leucopus*, palatal view of skull showing reduced premaxilla (premaxilla black).

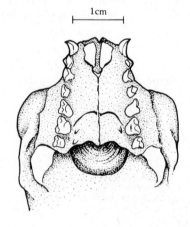

canines are subequal in size in males and females (e.g. *Hylobates*, the gibbon, many of the prosimians, and the South American owl monkey *Aotus*).

Distal to the canines are the premolars. Premolars have only one cusp in their simplest form. Around the base of the cusp a thickening of enamel is occasionally formed. This structure is called a *cingulum* (*cingulum* is the Latin word for "band"). A *cingulum* is found on many premolars and molars and sometimes also on incisors and canines. Premolars often have two cusps, one situated buccally and one lingually. On upper premolars these cusps are named the paracone (buccal) and the protocone (lingual). A single-cusped premolar has only a paracone. Names of the lower premolar (and molar) cusps end with the suffix -id, thus designating the tooth clearly as being of the lower dentition. Lower premolars of primitive primates have a distal basin consisting of an unstructured enamel thickening that may bear one or two cusps. The anterior part of the lower premolar or primitive primates has two cusps, the protoconid (buccal) and the metaconid (lingual). The protoconid is usually the highest and dominating cusp. In higher primates, lower premolars usually have only the two anterior cusps, the protoconid and metaconid and lack the distal basin.

In some cases the distal premolars resemble the molars, and the premolar series grade morphologically into the molar shape. For example, in extant galagos the fourth premolar is very much like the first molar.

The molars are the most distally positioned type of teeth. These are the most complicated teeth. Cusps, grooves, fissures, and ridges, in some cases wrinkles (e.g. the orangutan) give structure to the occlusal enamel. In primates the simplest upper molars have three main cusps. They are called cones. Smaller cusps that complicate the structure of the occlusal surface are commonly called conules, and they form within the range of ridges that connect the main cusps with each other. Two of the three main cusps are located on the buccal edge of the molar (the paracone and metacone); see Figure 3–15. The third main cusp lies lingually (the protocone). A fourth cusp, the hypocone, is present on the upper molars of most primates. This cusp occurs on the distal-lingual occlusal surface. One root is located under each of these three main cusps. Thus, all the upper molars have three roots, one inside and two outside. Each root is set into the maxillary bone, and the bony alveoli—the tooth sockets—are perfect negative images of the roots, the root dentine only covered by a thin layer of cementum, connected in turn to the alveolar bone by the peridontal membrane.

The triangular arrangement of the cusps and roots in upper molars is called the "trigon" or "trigonum." This type of tooth is also called "tritubercular" and was taken as the basic tooth type for upper molars from which all teeth of extant mammals can be derived following Cope and Osborn's tritubercular theory (1888).

The lower molars consist of two parts. Both parts have three main cusps and thus, lower molars of early primates have six main cusps. Also, the three cusps of the two components that are joined together to form the basic pattern of lower molars are combined in triangles. The triangle in the front has two cusps at the lingual side (paraconid and metaconid) and one bucally (protoconid)—just the inverse of the upper triangle. This mesial (anterior) triangle is called the "trigonid" and is believed to be phylogenetically older than the back part, the so-called talonid or heel. The talonid consists of one buccal cusp (hypoconid), one lingual cusp (entoconid), and one distal cusp (hypoconulid); see Figure 3–15. The triangular parts are fused together to compose an elongated molar tooth, and in early forms the trigonid level lies considerably higher than the talonid level. Among extant primates this condition is still expressed to some extent in *Tarsius*. Lower molars have two roots anchoring the tooth into the mandible, one mesial (anteriorly) under the trigonid and one distally under the talonid. In most extant primates the two regions of lower molars—the trigonid and the talonid—have their occlusal surfaces at almost the same level.

Dental Function

There is a functional and morphological dependency between the maxillary and mandibular tooth rows. The structures (cusps, crests, and grooves) of the occlusal surfaces of all the corresponding upper and lower teeth fit into each other (Kay and Hiiemae, 1974). The morphology of the teeth varies in relationship to their function. The following discussion considers a number of primate dental morphologies and their functions.

The primary function of incisors for primates is the preparation of food items for mastication. Hylander (1975) has documented a relationship between incisor size and the diets of higher primates. Higher primates that feed primarily on large objects (i.e. fruit) often have larger incisors than primates that eat small objects (i.e. seeds, leaves, grasses) because it is necessary for them to reduce the size of their food so that they can chew it between their molars. *Callithrix* and *Cebuella*—two small New World monkeys—have relatively long lower incisors that they use to gouge holes in bark to induce gum and sap flow (Coimbra-Filho and Mittermeier, 1978). As described previously, members of Prosimii, with the exception of *Tarsius*, have specialized lower incisors that function as a grooming comb.

The relationship between canine morphology variation and function is less well understood in primates than this relationship for incisors and molars. Many primates do use their canines for ripping and slicing food items. Canines are also used in aggressive encounters by many primates, both as a weapon and as a threat object. *Callithrix*, *Cebuella*, and *Phaner furcifer* use their canines to gouge holes in bark (along with their incisors).

Morphological variation of premolars, like that of canines, is not clearly understood in relationship to function. The anterior lower premolar of many primates with large canines acts as a hone to sharpen the distal crest of the upper canines. The most specialized examples of this condition are seen in such Old World monkeys as baboons and macaques where the P_3 is elongated mesially-distally with a long surface on its buccal surface that hones the upper canine. *Galago elegantulus* has a P_2 that is caniniform. This specialized premolar is used to gouge holes in bark (Charles-Dominique, 1977).

As mentioned previously, the molars are the most complex teeth morphologically. Reflecting the close ties between morphology and function, the molars are also the most complex teeth functionally. The primary function of the molars is the mastication and preparation of food for digestion. Mastication for primates is a fairly stereotypic behavior consisting of two phases (Kay and Hiiemae, 1974). Phase I consists of an upward and antero-lingual movement of the lower molars across the upper molars until the protocone is in the deepest point of the talonid basin (centric occlusion). Phase II follows immediately with the lower molar continuing its antero-lingual movement downwards until the molars are no longer touching. During this chew cycle, food is being prepared for digestion in three ways. Firstly, food is being sheared between the crests that connect the cusps on both the upper and lower molars. Secondly, food is being crushed between the lingual surfaces of the buccal cusps of the lower molars and the buccal surfaces of the lingual cusps of the upper molars during Phase I. Thirdly, food is being ground between the lingual surfaces of the buccal cusps of the lower molars and the buccal surfaces of the lingual cusps of the upper molars during Phase II. For a complete discussion of shearing, crushing, and grinding, see Kay and Hiiemae (1974), and Kay (1975, 1978).

While Phase I and Phase II mastication is a constant feature of primate chewing, certain structures of the molars are elaborated in different primates to increase the efficiency of mastication. For primates that eat primarily leaves or insects, it is more important to reduce food particle size during mastication than for primates that eat primarily fruit. The reason for this is that the digestibility of cellulose (leaves) and chitin (insects) is increased by reducing food particle size, whereas the digestibility of fruit is not increased by reducing food particle size. Leave-eating and insect-eating primates have longer shearing crests than do fruit-eating primates. This increases shearing capacity in both Phase I and Phase II mastication (Kay, 1975).

Within the Order Primates there are several trends modifying the number of cusps that combine to form the occlusal surfaces of molar teeth. Almost certainly *Tarsius* is the prosimian genus that now exhibits an occlusal morphology that most nearly resembles the original primitive eutherian molar pattern described above, where there are three major cusps in upper and six cusps in lower molars.

The most common trend affecting structure of teeth in primates is the addition of the hypoconid to upper molars and the reduction or loss of the anterior cusps of the trigonid part of lower molars. The hypocone is usually derived from the *cingulum*. Hypocone, however, is only a position name, for it is known to have arisen independently on repeated occasions. In those cases where the hypocone originates from the *cingulum* it is called a "true" hypocone. If a cusp in the same position develops by splitting of the protocone into two cusps, the new cusp is called a pseudohypocone, a term first used by Stehlin (1916). The subject of occurrence of hypocones among primates has also been discussed in detail by Gregory and Hellman (1926). In the North American early Eocene primate lineage running from *Pelycodus* to its descendant *Northarctus*, a pseudohypocone has arisen by splitting of the protocone. Among extant prosimians one can find a number of transitional stages between the three cusped type upper molar and the four cusped upper molar. In all cases, however, the hypocone, be it small or large, is derived from the *cingulum*.

In lower molars the general trend is toward reduction of the anterior or mesiolingual cusp, the paraconid. Also the most posterior or "third" cusp of the talonid—the hypoconulid— is reduced on the M_1 and M_2 of many extant primates.

Tupaiiformes

Genus Tupaia

The dental formula of all members of Infraorder Tupaiiformes is

$$\frac{2.1.3.3.}{3.1.3.3.}$$

Upper incisors. The upper incisors are of cylindrical shape. The central pair projects further out than the lateral pair and also has larger diameters. There are large diastemata between the central incisors and the central and lateral incisors.

Lower incisors. The two middle pairs of lower incisors are comparatively long and slightly spatulate. The lateral pair of incisors is, as in the upper dentition, smaller in length and diameter than the two other pairs and cylindrical in shape. Upper and lower incisors do not have real bite-contact but function more or less like a pair of pinchers. The lower incisors are tilted far forward and are implanted almost in line with the long axis of the mandibular ramus. Functionally, the three pairs of lower incisors make up a six-toothed comb that is used by the animals to comb their fur.

Upper canines. A wide diastema separates the upper canine from the lateral incisor. The canine morphology is only slightly different from that of the lateral incisor when unworn and is essentially like it when the point is worn off. The canine is slightly more pointed than the lateral incisor and curves somewhat backward.

Figure 3–6 *Tupaia* dentition (canine stippled, premolars black).

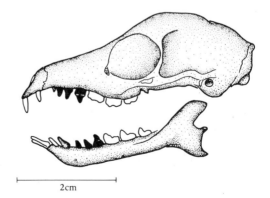

Lower canine. The lower canine is about as much separated from the lateral incisor as the latter is separated from the middle incisor. The lower canine is larger in diameter when compared to the first two incisors but about equal in length; it is caniniform with a backward directed apex. The lower canine is larger than the upper canine; it is implanted procumbently into the mandible.

Upper premolars. A small gap separates the first simply-shaped upper premolar from the canine. It is a low single-cusped tooth, also separated from the second premolar by a very small gap. Second and third premolars are increasingly larger in occlusal surface area and are triangular. These teeth have a single buccal cusp (paracone). This is much higher and more dagger-like in P^3 and P^4 than are the two buccal cusps on each of the three molars.

Lower premolars. The three lower premolars are different from each other in shape and projection height. The first lower premolar is much smaller than the canine (less than half the canine size), is separated from the canine by a small gap, is still somewhat procumbent, and is pointed with one cusp. The first and second premolars are separated from each other by a small diastema. When unworn, the second projects nearly twice as high as the first premolar. P_3 has two roots (the P_2 has only one) and one pointed cusp. Some individuals have a *cingulum* on the inner side of this tooth that extends to the end of the tooth, forming a small talonidlike extension. Following behind the second premolar is the two-rooted and two-cusped metaconid and protoconid third premolar. Its shape is transitional into that of the molars. At the distal end of the tooth, there is a small talonid. The level of the talonid is considerably lower than the apex of the protoconid, and the metaconid is lower than the protoconid but higher than the talonid. No distinctive cusps can be recognized on the small talonid, which juts back under a slightly forward-projecting paraconid of the adjoining M_1.

Upper molars. M^1 is the largest of the three. The first and second molars have a large inner *cingulum* that is especially marked in M^1 where it bulges out at the inside of the tooth, giving the occlusal surface a quadrangular shape. There is often a hypocone, derived from this *cingulum*, on M^1. All three upper molars have three cusps and three roots. M^2 is smaller than M^1 and M^3 is very small, having only about a third of the occlusal surface of M^2.

Lower molars. M_1 is only slightly larger than M_2. The trigonid occlusal surface is distinctly higher than the occlusal surface of the talonid. The protoconid is the highest cusp in unworn molars. An enamel protrusion on the front of M_2 extends under the backward-projecting small hypoconulid of M_1 like a strut. The same type of "interlocking" occurs between M_2 and M_3. In M_1 and M_2 the hypoconulid is crowded inside toward the entoconid and is separated from the hypoconid.

Lemuriformes and Lorisiformes

The dental formula of Lemuridae and Lorisiformes is identical with one exception: in *Lepilemur* the upper incisors are totally lost. In both Lemuridae (including *Lepilemur*) and Lorisiformes the lower incisors are procumbent, forming a toothcomb with the canines. The dental formula of Lemuridae (except *Lepilemur*) and Lorisiformes is

$$\frac{2.\ 1.\ 3.\ 3.}{2.\ 1.\ 3.\ 3.},$$

and the dental formula of *Lepilemur* is

$$\frac{0.\ 1.\ 3.\ 3.}{2.\ 1.\ 3.\ 3.}.$$

In the milk-dentition of *Lepilemur* small upper incisors are present, so that the deciduous dental formula of *Lepilemur* is:

$$\frac{2i.\ 1c.\ 3m.}{2i.\ 1c.\ 3m.}.$$

Genus *Lemur*

Lower incisors. These teeth are procumbent and are nearly parallel to the long axis of the mandible (Figure 3–7), the crown forming an angle with the root. There is no contact at all between upper and lower incisors. The lower incisors have long crowns, are very narrow and flat, and are implanted close to each other.

Figure 3–7
Lemur, maxillary dentition, (canine stippled, premolars black). For mandibular dentition, compare with Figure 3–2.

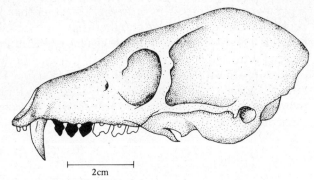

Upper incisors. The upper incisors are reduced to small and short cylindrical teeth. The central pair is somewhat smaller than the lateral pair. These two pairs of incisors are implanted rather laterally in the premaxilla and are thus separated by a wide gap. A result of this is that during occlusion with jaws aligned, the entire lower toothcomb (I_1. I_2. C) is positioned between the central upper incisors.

Lower canine. In the mandible the canines are also procumbent and are included in the toothcomb. Their crowns are nearly as long as those of the incisors but slightly more bladelike, or, to express this differently, they are somewhat broader than the latter. Also, the tooth crown tilts slightly inward toward the midline and is not aligned with its root.

Upper canines. In both sexes the upper canines are daggerlike, long, pointed, and bent backwards. These canines are positioned closely to the lateral incisor but separated from the first premolar by a wide diastema.

Lower premolars. The first lower premolar has one cusp and at the back a short heel. This tooth is shaped like a canine compressed from side to side and having one high pointed cusp. The entire height of the tooth may not exceed its greatest length at its base. A small gap separates the first premolar from the canine, and a gap of approximately the same size separates the former from the second premolar. The second premolar is considerably smaller than the first premolar and is positioned closely to the third premolar. The shape of the second premolar resembles that of the first premolar; it has, however, only about half the length of the latter; it also has a heel that is somewhat broader than the heel of the first premolar. Three cusps can readily be identified in the trigonid region of the third premolar. The paraconid is not as high as the protoconid and the metaconid, the protoconid being somewhat higher than the metaconid. No distinctive cusp can be recognized on the talonid basin, which consists of a central groove surrounded by a continuous ridge. The talonid of the third premolar juts under the front end of the first lower molar with its hind end.

Upper premolars. A comparatively wide diastema separates the first upper premolar from the canine. This tooth consists of one prominent cusp, the paracone, with a steep cutting edge on its anterior and posterior slopes, resulting in a pointed tooth. From the lateral aspect the second and third premolars look like slightly larger first premolars. First and second upper premolars are separated by a small diastema in some individuals. An additional cusp on the lingual aspect of the tooth, the protocone, enlarges the occlusal surface of the second premolar considerably as compared to the first premolar. In the third premolar the protocone is larger and more distinctive than the second premolar. The upper premolars gradually integrate morphologically into the molar morphology.

Lower molars. The first lower molar does not have a paraconid. In the trigonid area the protoconid is the most prominent cusp. A crest connects it with the somewhat lower metaconid. There is a low ridge at the front end of the tooth, and together with these two cusps this ridge surrounds the slightly inward shifted trigonid groove. The occlusal surface of the talonid is not much lower than that of the trigonid. In the talonid area the hypoconid is the highest and most distinctive cusp. The hypoconid is incorporated into a ridge that encircles the entire talonid area. Other cusps are not easily recognized. A distinctive talonid groove is surrounded by this crest and mesially by the trigonid elevation. The first molar is slightly larger than or equal in occlusal area to the second molar. Seen from the side it projects somewhat higher than the second molar and somewhat lower than the third molar. The third molar is considerably smaller than the two preceeding teeth. Also the third molar's crown pattern is simplified to an oblique crest that connects the metaconid with the protoconid, and the talonid is simply surrounded by crests.

Upper molars. The protocone is large in the first two molars and somewhat forward positioned. A distinctive *cingulum* at the inside holds two cusps in the first molar: in the front the pericone and at the back a somewhat smaller hypocone. This additional anterior cone within the inner *cingulum*—the pericone (Stehlin, 1916)—is much bigger than the hypocone in *Lemur mongoz* and *L. macaco* (Remane, 1860). Swindler (1976) calls it "protostyle." Also in the second molar the pericone is well developed, whereas the hypocone is usually absent. The first and second molar are subequal in size, the latter somewhat smaller in occlusal surface size and slightly lower in its projection above the alveolar margin. The third molar is considerably smaller than the second and simply formed with three cusps.

In Indriidae the deciduous dental formula is different from the adult dental formula. This is because a small canine is found in the deciduous dentition that is not replaced in the permanent set of teeth. (For discussion, see Remane, 1960; Bennejeant, 1935; and Friant, 1935). The lateral milk incisor is replaced by a permanent lateral incisor. This interpretation of the indriid dental formula is based on the evaluation of extant indriid dentitions, both deciduous and permanent. For an alternative interpretation of the indriid dental formulas, see Schwartz (1974). In Indriidae the two dental formulae are:

$$\text{milk dentition} \quad \frac{2i \cdot 1c \cdot 2m}{2i \cdot 1c \cdot 2m}$$

$$\text{permanent dentition} \quad \frac{2 \cdot 1 \cdot 2 \cdot 3}{2 \cdot 0 \cdot 2 \cdot 3}$$

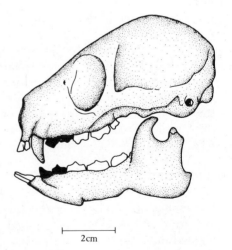

Figure 3–8 Indriid (*Propithecus*) dentition (canine stippled, premolars black).

2cm

In the permanent dentition the lower incisors are implanted at a procumbent angle, but the procumbency of the indriid toothcomb is less pronounced than the procumbency of the lorisoid and lemurid toothcomb. The premolars are shaped like canines and are not incorporated into the toothcomb.

In addition to the detailed description of tooth morphology in genus *Lemur* already given, the following is a more generalized description of the lemuroid and lorisoid dentition. This does not apply to the lemuroid genus, *Daubentonia* (discussed hereafter), which has the the most specialized dentition of all recent primates.

Cheirogaleinae, Indriidae, Lorisidae

Lower incisors. These are more or less parallel to the long axis of the mandible (procumbent) thus forming a toothcomb together with the similarly shaped lower canine. (Lower canines are lost in species of the genera *Avahi, Propithecus,* and *Indri*). These animals are called "toothcomb" prosimians.

Upper incisors. These teeth are generally small, having lost the functional bite contact with the lower incisors and are lost in *Lepilemur*. In genera *Propithecus* and *Indri*, on the contrary, the central upper incisors are still relatively large.

Lower canines. The lower canines are shaped like lower incisors and are corporated into the toothcomb with the exception of the three indriid genera *Avahi, Propithecus, Indri.* The lower canine is lost in the permanent dentition of Indriidae.

Upper canines. These teeth are always large, projecting, pointed teeth. They are curved distally (recurved) in Cheirogaleinae and Indriidae but are rather straight and peglike in Lorisidae.

Lower premolars. In all the prosimian dentitions summarized here, the first lower premolar has adopted the shape of a canine, thus replacing it functionally. It can easily be recognized as a premolar because, when upper and lower jaws are occluded, its position falls behind the upper canine, whereas a true lower canine occludes in front of the upper canine. The lower premolars following to the rear of the procumbent canine are usually blade-like, with only one cusp. The back premolar is often molariform, as can be seen in Lorisidae but most obvious in the galagos.

Upper premolars. The first two premolars are usually of simple design with only one cusp, whereas the third is commonly equipped with three cusps and resembles the molars. Especially in genus *Galago*, the premolars resemble molars, and the last upper premolar has four cusps, with the hypocone present.

Species of the cheirogaleine genus *Phaner*, as well as subgenus (*Euoticus*) among the Galaginae, have in common a peculiar arrangement of upper canine and first upper premolar. The canine is pointed and projecting, as canines usually are, but the somewhat shorter second premolar is shaped like a canine as well. The lower caniniform premolar fits in between the upper canine and the upper caniniform premolar (P^2) when jaws are closed. This is apparently an adaptation for gouging bark, which causes gum (an important part of the diet these animals) to ooze from the tree.

Lower molars. Usually the paraconid is missing, and the level of the talonid area is the same as that of the trigonid. The first molar is commonly the largest, and the second molar is only slightly smaller than the first. The third molar is usually the smallest of the three. This condition is especially pronounced in Lorisidae (except *Arctocebus*), where the occlusal area of the third molar is about half the size of that of M_2. In Indriidae and *Arctocebus* the third lower molar is subequal in size to the second and has a hypoconulid, which is missing in the first and second lower molars.

Upper molars. First and second molars have a hypocone except in genera *Cheirogaleus, Microcebus, Mirza* and *Perodicticus* that have three cusps on all molars. The third molar is small and always has only three cusps.

Daubentoniidae

The family Daubentoniidae with one extant genus and species, *Daubentonia madagascariensis*, is characterized by the most highly specialized dentition among all living primates. The teeth of *Daubentonia*—the aye-aye—are reminiscent of those of rodents, which led early naturalists to describe and classify the aye-aye as belonging to the mammalian Order Rodentia.

This peculiar tooth structure of *Daubentonia* also influences the outline and shape of the entire skull, contributing to the rodentlike appearance. The anterior dentition of *Daubentonia* not only resembles that of rodents in the number and shape of the teeth

Teeth

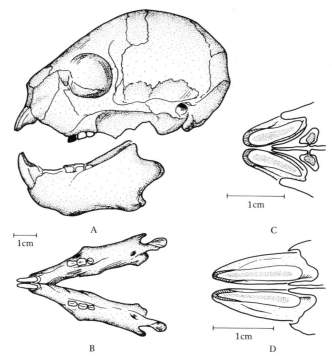

Figure 3–9 (A) *Daubentonia* maxillary dentition (premolar black); (B) mandibular dentition, occlusal view; (C) upper incisors, enlarged to show the enamel-covered front (enamel-striped) of the two incisors; (D) lower incisors enlarged, showing enamel.

but also in the distribution of the histological material that makes up the tooth. Enamel, if present at all, is very thin except on the anterior aspect of the front teeth where it is thick (see Figure 3–9C). It seems that the wear on these front teeth is governed by the hardness of the anterior layer of enamel. Moreover, it is quite unlikely that the lower gnawing teeth of the aye-aye are the canines, as has been suggested by some (e.g. Winge, 1895; Gregory, 1922). There can be no doubt that the rodentlike upper front teeth are implanted in the premaxilla, and thus incisors, as can be seen in subadult specimens (see Figure 3–9). *Daubentonia* has, in both maxilla and mandible, two deciduous incisors, one of which is not replaced in the permanent dentition. The dental formulae are:

$$\text{milk dentition} \quad \frac{2i \,.\, 1c \,.\, 2m}{2i \,.\, 0 \,.\, 2m}$$

permanent dentition $\frac{1.0.1.3}{1.0.0.3}$

Thus, a total of six deciduous teeth are not replaced in the permanent dentition. For an alternative and less widely accepted interpretation of *Daubentonia*'s dental formulae, see Tattersall and Schwartz (1974).

As stated already, the chisel-like shape of the aye-aye's front teeth as seen from the side is, in part, a product of the anterior enamel layer on these teeth that prevents the front side from being worn down at the same rate as the surface of the tooth covered only by dentine; compare Figure 3–9c,d and see Remane (1960).

The roots of the front teeth run far behind the molars in both upper and lower jaws and, like those of rodents, grow continuously. In the upper jaw the incisor root goes back to the root of M^2. Lower incisor roots extend even further, for they reach far beyond the roots of the three molars and enter into the coronoid process.

A very small tooth with only one root represents the upper premolars. Its occlusal surface is uncomplicated with only one not very prominent cusp. The three upper molars are much larger in occlusal surface area than this premolar. The third molar has almost as much surface area as do the first and second molars. All three molars in both upper and lower jaws have comparatively low, structureless cusps.

Tarsiidae

Reduction of tooth number can be regarded as a progressive feature in primates. Genus *Tarsius* has lost an additional pair of lower

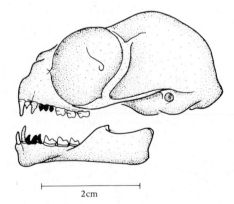

Figure 3–10 *Tarsius* dentition (canines stippled, premolars black).

incisors when compared to Lemuridae. The tooth morphology of tarsiers, however, resembles most closely the occlusal structure of a particular group of early primates, the Anaptomorphidae, and can therefore be regarded as primitive. The *Tarsius* dental formula is

$$\frac{2.1.3.3}{1.1.3.3}$$

Schwartz (1978) has recently suggested an alternative dental formula for *Tarsius*; however, it has not gained wide acceptance.

Lower incisors. Only two small (one on each side), cone-shaped incisors are present. These are not implanted in a procumbent position (as in toothcomb prosimians) but face directly upward.

Upper incisors. The inner pair of cone-shaped incisors is much larger than the lateral pair. Both pairs have the same simple and basic conical morphology.

Lower canines. The lower canines are larger than the incisors and are shaped like typical canines.

Upper canines. The upper canines are caniniform as well. They do not, however, project much higher than the lateral incisors and are less high than the central incisors. Nevertheless, at the base the diameter is greater than in I^1. There is only a very small gap between the lateral incisors and the canines.

Lower premolars. The simple, two-cusped premolars increase in size successively from the first to the third. The lingual trigonid cusp is very small. The first premolar is only about half the size of the canine and even the third is smaller than the canine.

Upper premolars. Of the upper premolars the first is smaller than the canine and it is unicuspid. P^3 and P^4 (second and third premolars) have two cusps, with the cusp at the inside smaller than the outside cusp. The third premolar is the largest of the three.

Lower molars. In tarsiers the trigonid of the lower molars retains three very pointed cusps. The occlusal surface of the talonid is lower than that of the trigonid. Hypoconid and entoconid are well developed, whereas the hypoconulid is comparatively small. The third molar is somewhat lower than the first and second molar, but its occlusal surface is just as large.

Upper molars. All three upper molars are approximately equal in occlusal surface size but decrease in height from the first to the last molar. Almost as in Insectivora or Chiroptera (bats), the three main cusps, protocone, paracone, and metacone are very pointed and prominent on the upper molars. Paracone and metacone are higher than the protocone. A well-developed *cingulum* on the lingual distal surface of the molars shows a thickening that represents the hypocone. This *cingulum* and the tiny hypocone are usually very small and occasionally absent on M^3.

Anthropoidea

Tooth number as well as tooth morphology are more uniform among higher primates than in extant Prosimii. Common to all higher primates is the loss of the paraconid on lower molars, the nearly equal height of the occlusal surface of trigonid and talonid areas in the mandibular molars, and the presence of four cusps in all maxillary molars (this number has been secondarily reduced in Callitrichidae). The number of premolars is distinctive for the two major groups of higher primates: all higher primates of the Old World have two premolars in the upper and lower jaw

$$\frac{2.\ 1.\ 2.\ 3.}{2.\ 1.\ 2.\ 3.}$$

those of the New World have three in both jaws

$$\frac{2.\ 1.\ 3.\ 3.}{2.\ 1.\ 3.\ 3.}\ \text{Cebidae}$$

and

$$\frac{2.\ 1.\ 3.\ 2.}{2.\ 1.\ 3.\ 2.}\ \text{Callitrichidae.}$$

Cercopithecoidea: Old World Monkeys

All members of Cercopithecoidea have a dental formula of

$$\frac{2.\ 1.\ 2.\ 3.}{2.\ 1.\ 2.\ 3.}.$$

Their molar morphology is very consistent within the superfamily. In Cercopithecoidea, the main cusps of the molars are positioned in two pairs inside and outside and are directly adjacent to each other situated as at the corners of a square. They are usually connected by straight but not always prominent cross ridges that run at right angles to the long axis of the teeth. Individual cusps are, when unworn, usually more prominent in crown height than are those of typical hominoid molars. This symmetrical position of the molar cusps and their connection by ridges across the tooth form a characteristic cusp pattern which is called "bilophodont"; see Figure 3–15. This molar pattern is not only found in cercopithecoid primates but also in pigs, tapirs, rhinoceroses and, somewhat modified, in some marsupials. This pattern is typically found in the two first molars of the lower jaw of cercopithecoid monkeys and in all three upper cercopithecoid molars. The third lower molar of cercopithecoid monkeys usually has an additional heel-like elongation of the hind end of the tooth. The hypoconulid is missing in cercopithecoid M_1 and M_2. The hypoconulid is also usually missing in third molars of members of the genera *Cercopithecus* and *Erythrocebus* (patas monkey). A controversy exists as to whether earliest cercopithecoids had enlarged hypoconulids on M_3 or whether primitively M_3 more closely resembled M_2.

Figure 3–11 Dentition of Old World cercopithecoids, *Cercopithecus* spec. (canines stippled, premolars black).

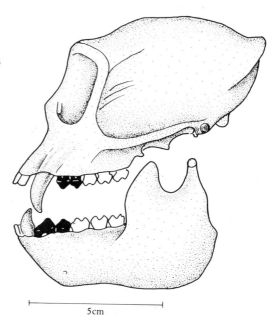

5cm

Lower incisors. The lower incisors of Old World monkeys are usually spatulate and subequal in size. The central pair tends to be broader than the lateral pair of cercopithecines; in colobines the lateral incisors tend to be broader. The incisors are mostly implanted in a straight upward-directed way, thus being orthally oriented rather than procumbent.

Upper incisors. The upper incisors are spatulate and the central pair is always broader than the lateral pair. Typically, the upper incisors are not implanted as orthally as the lower ones but slope forward. Frequently, the distal (lateral) pair is tilted mesially. This condition is often very pronounced in baboons. In individual cases it can result in so-called "crowding": a slight overlapping of the lateral tooth edges at the tooth tips. The relative size difference of the upper incisor pairs is much more pronounced in Cercopithecinae than in Colobinae.

Lower canines. There is a marked sexual dimorphism of the canines in many of the living representatives of the superfamily Cercopithecoidea. It is most marked, however, in the genus *Papio*. The canines are pointed, powerful teeth that curve backward and have a heellike extension on their hind end.

The lower canines of baboons have also a marked groove running up and down the anterior aspect or face of the tooth crown. This condition is more pronounced in males than in females. In addition, a similar anterior groove is also found in the upper canines of cercopithecoids. This groove is best developed in baboons and macaques where it not only extends up and down the conical crown of the tooth but also runs down the entire length of the root.

Upper canines. The upper canines usually project to a considerably higher crown length than the lowers. The upper canines are dagger-like and slightly curved backwards. Together with the lower canines and the highly specialized first (P_3) lower premolar, the upper canines function as a very efficient shearing device. The upper canine and the lower first premolar also sharpen each other constantly during function (see Every, 1970). The hind edge of the upper canine is worn against the front, elongated edge of the two-rooted first lower premolar. This honing mechanism sharpens the shearing edges of these teeth to blade-like cutting edges, a phenomenon again most pronounced in male baboons.

Lower premolars. The first lower premolar (P_3) is highly specialized in all the genera of the superfamily Cercopithecoidea; it

Figure 3-12 Honing lower premolar.

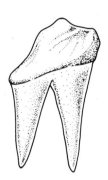

commonly has two roots, the anterior of which is considerably shorter than the distal one (Figure 3-12). The enamel forms a sharp cutting edge extending down the sole cusp (located above the distal root) to the front root. The enamel coating also extends far down onto the front root. The front root is implanted more buccally than is the hind root. The result of this orientation is an outward tilt of the cutting edge. Usually, the front root has a larger diameter than the distal one. This type of premolar with the long forward sloping cutting edge has been called "sectorial," but perhaps "honing" is a better term. This premolar is typically developed in all the representatives of the superfamily Cercopithecoidea.

P_4 often has four cusps, the anterior pair of which is bigger than the rear pair. Instead of a paraconid we usually find a cross ridge on the mesial end of the crown. This crest bends distal-lingually to run to the metaconid and distal-buccally to run to the protoconid. These cusps are usually connected with each other by an additional cross ridge. The protoconid tends to be slightly higher than the metaconid. From both of these cusps a ridge slopes distally on each side of the tooth and surrounds the comparatively small talonid area. The occlusal surface of the heel is positioned below that of the trigonid, and the talonid basin is well marked.

Upper premolars. The cercopithecoid upper premolars have comparatively simple structures. Both have two principal cusps (protocone and paracone) that are connected by a cross ridge. The paracone is usually higher than the protocone. P^4 has a larger occlusal surface than P^3. However, the height of the crowns of the two teeth are equal. Both the mesial and the distal ends are outlined by ridges on the unworn occlusal surface, with the hind part larger and forming a small heel in P^4 (less pronounced in P^3). The upper premolars usually have three roots, two positioned under the outside of the crown and one under its inside.

Lower molars. All three lower molars of members of this superfamily show the typical bilophodont cusp patterns: four major cusps (protoconid, metaconid, entoconid, hypoconid) are arranged opposite each other (inside and outside) and are connected by cross ridges. In the trigonid area the paraconid is lost and usually is replaced by a ridge sloping down across the mesial end of the tooth. The level of this ridge or crest is as low as the bases of the cusps. The lingual cusps are usually higher than the buccal cusps. The first molar is commonly the smallest of the three, the third one the largest. In *Cercopithecus* and *Erythrocebus* and some of the Colobinae, the size difference of the three lower molars is less pronounced than in, for example, the genera *Papio* or *Macaca*. Lower molars have two main roots, a proximal one and a distal one; each may divide in two toward its apex.

Upper molars. The upper molars are very similar in morphology to the lower molars. M_1 through M_3 have four cusps, paracone, protocone, metacone, and hypocone. The paracone and metacone (the buccal cusps) usually project higher than the inside cusps in both unworn and worn teeth. In some of the genera of small body size, *Cercopithecus (Miopithecus) talapoin*, for example, the M^3 tends to be the smallest tooth, with only three cusps. The size differences between the upper molars are usually less pronounced than are those between the lower molars. Upper molars have three roots, two bucally and one lingually.

Ceboidea: New World Monkeys

Cebidae

Upper incisors. In general, ceboids have upper incisors that are spatulate, with the central pair much broader and also longer than the laterals. There is a long diastema between the incisors and the large tusklike canines. *Cacajao, Chiropotes,* and *Pithecia* have a specialized front dentition, for both upper and lower incisors are implanted tilting forward. The forward tilt of the incisors in these animals is more pronounced in the upper incisors than in the mandibular incisors. The peculiar species *Callimico goeldii* also has forward tilting upper incisors but to a lesser degree than in *Cacajao, Chiropotes,* and *Pithecia.*

Lower incisors. The lower incisors are subequal in size, rather slender and long, and a small diastema can usually be detected between incisors and canines. Both upper and lower lateral incisors tend to have their crowns tilted inward. Usually only the internal upper incisors have bite contact with the lower incisors meeting them in an angle of about 90 degrees. Thus, they function rather like a pair of tweezers. There is a profound difference between the procumbency of *Cacajao, Chiropotes,* and *Pithecia* and prosimian front dentitions. Among those prosimians that have the most pronounced forward tilt of the lower incisors (plus canines), this tilt is mostly the result of an angulation between tooth and crown. In these prosimians and New World monkeys the entire tooth is straight and is implanted at a tilt in both the forward sloping mandible and the forward sloping maxillary bone.

Canines. The canines are caniniform, and in the forms with procumbent incisors the canine is not included in this "procumbent" tooth comb-like structure. In Cebidae, the canines show a marked sexual dimorphism being smaller in females than in males.

Upper premolars. Premolar morphology is not as uniform among Ceboidea as in Old World monkeys. The dental formula of all South American monkeys differ from those of the Old World monkeys in the retention of three premolars (Figure 3–13). The upper premolars are usually bicuspid (protocone and paracone); however, *Alouatta* often has a single cuspid P^2 (paracone). A third cusp (hypocone) is occasionally found on P^3 and P^4.

Lower premolars. In ceboid monkeys the anterior lower premolar (P_2) is enlarged and functions against the back of the upper canine as a honing mechanism. This structure is, however, not so highly specialized as the sectorial premolar found in Cer-

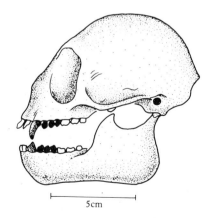

Figure 3–13 Dentition of New World cebids, *Ateles* spec. (canines stippled, premolars black).

5cm

copithecoidae. Cusp number often increases from P_2 to P_4, with P_2 often having a single cusp, P_3 two cusps, and P_4 three cusps (Swindler, 1976).

Upper molars. Molar morphology is fairly variable in cebids. M^1 and M^2 are usually square with four cusps (paracone, protocone, metacone, and hypocone). M^3 occasionally has four cusps but is commonly quite reduced in size and in cusp number. Mesostyles are common on the M^1 and M^2 for *Brachyteles*, *Lagothrix*, and *Alouatta*.

Lower molars. The lower molars usually have four cusps (protoconid, metaconid, hypoconid, and entoconid). In addition, *Ateles* also often has a hypoconulid (Swindler, 1976). A prominent crest connecting the protoconid and metaconid (protocristid) is quite common on the lower molars. Some cebids (e.g. *Brachyteles*, *Cebus*) have lower molars that appear to be similar to the bilophodont condition of the molars of Old World monkeys. This condition has suggested to some that bilophodont teeth have been derived more than once from the archaic primate molar pattern.

Callitrichidae

Upper incisors. The upper central incisors tend to be spatulated and much larger than the lateral incisors. A lingual *cingulum* is common on both I^1 and I^2.

Lower incisors. As discussed earlier, *Cebuella* and *Callithrix* have specialized lower incisors. They are elongated compared to the

Figure 3–14 New World callitrichid dentition, *Callithrix* spec. (canines stippled, premolars black).

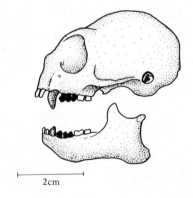

incisors of other callitrichids. In fact, these incisors are approximately as long as the lower canines.

Canines. The canines are caniniform and relatively robust. According to Wettstein (1963) the canines are longer in females than in males in these small New World monkeys.

Upper premolars. There is little variation in the upper premolars of callitrichids. These premolars are uniformly bicuspid, and the P^3 and P^4 often have a buccal and lingual *cingulum*.

Lower premolars. The lower premolars show a trend toward increased complexity from P_2 to P_4. P_2 is usually a single-cusped, slightly caniniform tooth. P_3 and P_4 are both bicuspid, and P_4 also often has a small distal heel.

Upper molars. Callitrichids are unique within extant primates in having only two molars; they have lost M^3. Callitrichids are also the only higher primates that normally has only three cusps on their upper molars (paracone, metacone, protocone). It is believed that this condition is not primitive but is a derived morphological state. Hypocones are present in low frequencies on the molars of *Saguinus* and *Callithrix* (Swindler, 1976).

Lower molars. The lower molars are almost square in shape. They usually have four cusps (protoconid, metaconid, entoconid, and hypoconid). As with the cebids, a strong protocristid connects the protoconid and metaconid.

Hominoidea

Gregory (1916) described the cusp and fissure pattern of the lower molars of the fossil hominoid genus *Dryopithecus*. He considered this pattern of the occlusal surface of molars to be typical of all hominoid primates, whether fossil or extant. This "*Dryopithecus*" molar pattern (in contrast to bilophodont) is of very important diagnostic value as a feature separating the two superfamilies of Old World primates, the Hominoidea from the Cercopithecoidea.

The typical "*Dryopithecus*-pattern" or "Y-5" has three buccal cusps and two lingual cusps in lower molars. The paraconid is lost. These five cusps are separated from each other by Y-shaped fissures; the two upper arms of the Y open buccally and embrace the

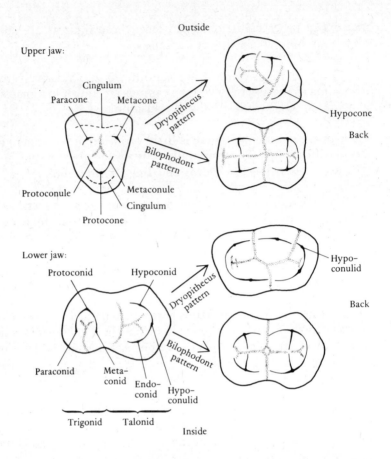

hypoconid, separating it from the protoconid (anteriorly) and from the hypoconulid (distally), and the long lower arm of the "Y" separates the two inside cusps, the entoconid from the metaconid (Figure 3–15). However, the *Dryopithecus* pattern is only occasionally found in this typical and more or less symmetrical configuration. The pattern undergoes numerous variations and simplifications and is often different even between the three molars of individual tooth rows (see Swindler, 1976). Among recent primates the pattern is most uniformly expressed in lower molars of gorillas (Remane, 1960).

In the upper molars of hominoid primates, we find the "*crista obliqua*," a ridge that connects metacone and protocone and that also delineates the hind edge of the original tricuspid upper molar, behind which the hypocone has been added. Typically, the small, additional cusp sometimes found between metacone and protocone in the original tricuspid upper molar, the metaconule, is incor-

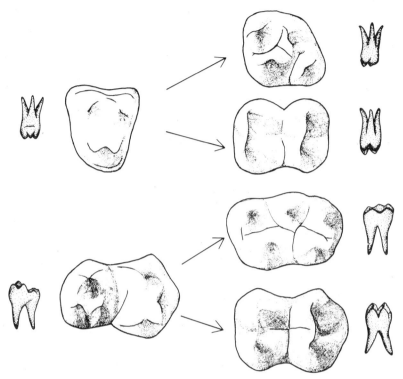

Figure 3–15 (*Facing pages*) Diagram of the development of the occlusal surface pattern in the molars of Old World monkeys and apes (ridges black, grooves stippled).

porated into the *crista obliqua*. This crest is a typical feature of hominoid upper molars and not of cercopithecoids. Also, the four cusps (when present) in hominoid upper molars are usually arranged in alternative positions (as if at the corners of a diamond) and not opposite to each other as in bilophodont upper molars of the Cercopithecoidea (see Figure 3–15).

Hylobatidae

Among gibbons there is little sexual dimorphism between males and females both in body and in canine size. Both females and males have long projecting, and distally recurved upper and lower canines. The first lower premolar of gibbons is somewhat enlarged in a mesial-lingual direction and functions together with the hind edge of the upper canine as a shearing device as was described for Cercopithecoidea.

Upper incisors. The central pair is somewhat broader than the lateral incisors. Both pairs are comparatively narrow.

Lower incisors. The two pairs are about equal in breadth and height. Their implantation in the rather shallow mandible is comparatively vertical when contrasted to pongid lower incisors in which the implantation is more procumbent.

Upper canines. The crowns of the upper canines are high and much broader across their base than lower canines. They are less recurved than the lowers and are much more daggerlike than the latter.

Lower canines. In both sexes the canines are large and pointed with a recurved tip.

Upper premolars. The upper premolars of gibbons are bicuspid. The paracone is the largest cusp on both P^3 and P^4.

Lower premolars. The first lower premolar (P_3) has one cusp (protoconid) and is elongated. P_3 has a heel-like enlargement (*cingulum*) distally. The second lower premolar (P_4) is also elongated but is always somewhat shorter than the first premolar. It has a paraconid, metaconid, and a heel-like distal extension that is formed by the *cingulum*.

Upper molars. Usually the second upper molar is the largest in occlusal surface area. In general, the hypocone is well developed in all upper molars, and their cusps are low.

Lower molars. The first lower molar is the smallest of the series, whereas M_2 and M_3 are nearly equal in size. Usually, all five cusps are present in the three pairs. The hypoconulid, however, can be missing in M_3. Generally the cusps are rather low and flat.

Pongidae

Upper incisors. The incisors are usually broad and spatulate. In the upper dentition the central incisors are broad at the cutting edge, with the lateral pair much smaller than the central pair (especially in *Pongo*, the orangutan).

Lower incisors. The lower incisors are subequal in size. There is a tendency toward a slightly procumbent implantation in pongid incisors, and procumbency is usually more marked in maxillary than in mandibular incisors.

Canines. The canines of apes are comparatively sturdy and tusk-like and not dagger-like as are the canines of the hylobatids. Gorillas and orangutans show distinct sexual dimorphism in the size of their canines. This condition is not as pronounced in chimpanzees. In chimpanzees (*Pan*) the canine height of males and females overlaps considerably. Consequently, the size of canines in chimpanzees cannot be used to distinguish all members of the two sexes (see Remane, 1960). There are usually large diastemata in the maxillary dentition. One is between upper incisors and canines and seems to correlate with the diastemata between the mandibular incisors and canines. A second is between lower canines and front premolars. Diastemata, however, are extremely variable features in pongids and do not always occur (see Remane, 1921, and Schultz, 1948).

Upper premolars. Both P^3 and P^4 are usually bicuspid with the paracone higher than the protocone. P^4 is usually somewhat larger than P^3. The upper premolars of pongids commonly have three roots, two situated buccally and one lingually.

Lower premolars. P_3 is a sectorial tooth (hone for upper canine) and usually has one cusp. P_4 is bicuspid with a short talonid area. This heel rarely has any distinctive cusps on its occlusal surface and is less than half as long as the anterior part of the tooth.

Upper molars. The upper molars of pongids follow the general description of that of Hominoidea given previously. M^1 is smaller than M^2. M^3 is occasionally larger than M^2, but it is also commonly the same size or slightly smaller than M^2.

Lower molars. The three lower molars have lost the paraconid but have the remaining five cusps. The protoconid and metaconid are positioned opposite each other, the protoconid buccally, the metaconid lingually. The metaconid and entoconid are separated by a conspicuous groove, the stem of the "Y-5." The two grooves that describe the arms of the upper half of the Y embrace the hypoconid between each other in the typical *Dryopithecus* pattern. As stated already, there are numerous modifications of this pattern, depending on variations in size and position of the cusps that are correlated with consequent changes in the arrangement of grooves. A reduction of the "Y" pattern to a simple cross pattern is frequently found in chimpanzees and in man. This can either come about through an enlargement of the hypoconid area, causing all grooves to meet or to cross in the center of the occlusal surface of the tooth so that the grooves form a five-armed star. Also, the reduction of the hypoconulid can result in a four-cusped lower molar with a simple right angle cross of the grooves. Size variation of the lower molars follows the same pattern as the uppers.

Hominidae

In most respects the dentition of hominids is similar to that of pongids. The following is a short discussion of some important differences that do occur between these dentitions. Typically, the incisors of extant hominids are implanted straight up and down. Both the upper and lower premolars are bicuspid, with buccal cusps larger than the lingual cusps.

One of the most characteristic features of the hominid dentition is the reduced size of the canines, which rarely project much beyond the crown height of the incisors or premolars. Further, diastemata rarely occur in either the maxillary or mandibular dental arcades.

In hominids the lower molars tend to have lost the hypoconulid and to have the remaining four cusps arranged in a cross (+) or star (★) pattern. The latter occurs in those lower molars that still have five cusps. Also the lengths and breadths of hominid molars are usually subequal. Thus, the occlusal surface of the molars is rather square. Third molars show a strong tendency toward size reduction and also a strong tendency towards reduction of occlusal morphology detail.

Finally, the dental arcade of modern hominids is different from that of all other primates. It is a more or less half-circle in accordance with the rearrangement of the morphology of the human face.

chapter 4

Skull

Primates show several major trends that are expressed in the structure and composition of the skull. These major trends are the enlargement and increasing complication of the brain and nervous system, the improvement of vision, and the relative reduction of those skeletal elements that are part of the nasal region. Among higher primates some of the functions of the dentition have gradually been taken over by the hand: for example, obtaining and preparing food. Also, the foramen magnum—the place of connection between skull and veterbral column—has a tendency to shift from the hind end of the skull toward the center of the skull base in successively more and more advanced groups (See Figure 4–1).

Looking at the primate skull from an ontogenetic and phylogenetic point of view, there are two different types of bone formation. In the skull there are both cartilage replacement bone and dermal bone. Cartilage replacement bone is pre-formed in cartilage; that is, the cartilage is replaced by bone during the ontogenetic development of the primate's body. All the elements of the body skeleton except the clavicle develop in this way. Within the skull the main parts of the skull base, the bony elements of the inner nose, the ear ossicles, and the cranial part of the hyoid apparatus are preformed in cartilage. The other type of bone development, which results in formation of the so-called dermal or intermembrane bones, begins with the connective tissue directly giving rise to bone within the skin. During ontogeny the connective

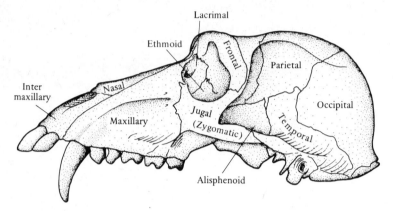

Figure 4–1 *Papio* skull.

tissue membrane is also replaced by bone. Most of the flat bones of the skull develop from dermal origins, including the bony elements that build the facial part of the brain case and, as already indicated, the clavicle. Clavicles are derived from a phylogenetically old element of the outer skeleton of primitive tetrapods. Within the Order Primates one finds a very distinctive difference between prosimians and anthropoidean primates in details of skull structure: In all prosimians except the highly distinctive East Asian group of the tarsiers, the eyes are surrounded from the back and the side by a bony ring. This ring consists of a part of the jugal and part of the frontal bone. In the Anthropoidea the eyes are almost completely enclosed by bony eye-sockets. Orbital walls of back and side are in the latter group composed of plates, extending out from the jugal, frontal, and behind from the sphenoid. The enormous eyes of *Tarsius* are, as in the Anthropoidea, enclosed by a bony socket. This is, however, only partly closed and leaves open an area to the sides and below. The socket is relatively larger than in platyrrhines and catarrhines. Also, in *Tarsius*, jugal, frontal, and a part of the sphenoid (alisphenoid) participate to form the postorbital closure. The roots of the molars enter into the orbital floor in *Tarsius*. (Haines, 1950). This postorbital closure is not as complete as in the Anthropoidea: An opening remains between the maxilla and the jugal (plus frontal) in the floor of the orbit of *Tarsius*.

In Anthropoidea postorbital closure is usually more complete than in *Tarsius* with one possible exception: The South American night monkey *Aotus*. This monkey resembles *Tarsius* because it retains a fairly large opening. Cartmill has recently observed that this opening is often nearly as large in *Homo* as in *Aotus*. Most extant anthropoideans show a much more reduced inferior orbital fissure. In *Aotus* the main part of the lateral closure is primarily formed by the zygomatic bone. In *Tarsius* alone the zygomatic bone contributes the lower part of the closure of the orbit, and the frontal bone forms the upper half of the lateral postorbital wall (Figure 4-2).

The lower jaw or mandible is a dermal bone consisting of two halves, the dentaries, which meet medially in a symphysis: A contact between adjacent bones. This symphysis fuses across and is obliterated early in life in all anthropoideans, but it remains unfused in nearly all living prosimians. In typical prosimians the mandibular rami are connected by tight ligaments and (with few exceptions) never fuse into one solid mandibular bone.

All lower teeth are located in the horizontal ramus of the mandible. Behind the tooth row the mandibular rami ascend vertically (ascending branch) and, in front, have an upward projection (coronoid process) for the attachment of part of the chewing musculature. At the back the mandibular rami run up to knobs that articulate the mandible with the skull, the mandibular condyles. The lower hind end of the mandible, the angle, which also provides a surface for the insertion of chewing muscles, is hooklike and elongated in all extant prosimians except the indriids and the aye-

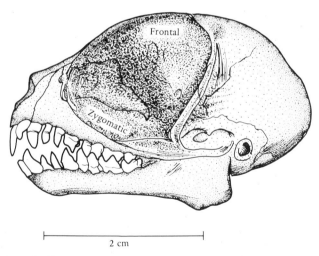

Figure 4-2 *Tarsius* skull.

aye *Daubentonia*. The angle of the mandible is rounded off in *Daubentonia*, indriids, and all higher primates.

The ascending rami differ in height in most prosimians on one hand and the majority of higher primates at the other. The articular condyles are positioned at the same level as the chewing surface of the tooth row in all prosimians, except the Intriidae and *Tarsius*, which resemble the anthropoidean condition in having the condyles high above the level of the teeth (See Figure3–8). However, in a small number of higher primates, the distance between the level of the mandibular articulation and that of the occlusal surface of the teeth is low, particularly in marmosets, in the howling monkeys—*Alouatta*—and in the gibbon—*Hylobates*. Seen from above or below, the mandible of prosimians is typically V-shaped and that of living Anthropoidea more or less U-shaped. The mandibular rami of most extant prosimians are more lightly built than is usually the case in higher primates. The facial skeleton is positioned in front of the prosimian brain case, giving most prosimian skulls a rather foxlike appearance. The brain of prosimians itself is ovoid and flat compared to that of monkeys. The inner part of the skull base (from front to back) formed by the sphenoid, pterygoid and occipital bones, is typically flattened in prosimians other than *Tarsius*. In most higher primates save for *Alouatta*, the inner part of the skull base is bent. The part of the skull base formed by the sphenoid and basioccipital is bent downward and thus forms an angle with the front part of the inner skull base, which is made up of the sphenoid and pterygoid (See Figure 2–10).

In prosimians the backwardly-directed position of the foramen magnum, the opening that meets the vertebral column, is a result of the flat skull base. Again, *Tarsius* is an exception among the prosimians. In *Tarsius* the foramen magnum is positioned centrally under the brain case. The exceptional appearance of the skull of *Tarsius* primarily derives from the enormous eyes of this animal. The skull of the majority of higher primates is characterized by having the orbits and facial skeleton partly tucked under the brain case rather than extending out in front of it. A partial exception to this rule is seen in baboons, whose snouts are large and long, a condition that formerly led to their being called *Cynomorpha*, which means "doglike."

Within the facial skeleton, upper teeth are rooted in the premaxillae and maxillae, both paired dermal bones. Whereas all three posterior types of teeth (canines, premolars, and molars) are rooted in the maxillary bone, only the incisors are located in the premaxilla. As long as the premaxillary-maxillary suture is visible, this condition provides a good criterion for distinguishing between incisors and the other teeth even when they are morphologically alike.

In some prosimians the upper incisors are reduced (e.g. *Lemur* or totally missing (e.g. *Lepilemur*), and in these cases the premaxilla also undergoes reduction. A narrow pair of membrane bones covers the nasal cavity from above. These elements are called the nasal bones. The maxillary bones and the lower margin and floor of the orbits are perforated by one or two openings—the so-called infraorbital foramina—through which the maxillary nerve penetrates. Behind the maxillary bones the paired palatine joins to form the hindpart of the bony (or hard) palate. The palatine bones are membrane bones, that is, formed by intramembranous ossification.

The nasal cavity is covered above by the nasal bones forming the bridge of the nose. The ethmoid on top and the vomer below fuse together to form the medially positioned nasal septum that separates the two nasopharyngeal fossae. The vomer rests on top of the palate and the maxillae and is contacted by the ethmoid and sphenoid from above and behind respectively. In living primates both the medial nasal septum and the anterior margins of the nasals and maxillae are extended forward by cartilage. These cartilaginous nasal wings are usually broad in South American monkeys, narrow in Old World monkeys and apes as well as in humans.

This difference led to the taxonomic distinction of these two groups into the infraorders Platyrrhini and Catarrhini. This distinction has been criticized because the breadth of the cartilaginous nasal wings is highly variable in primates and thus of minor taxonomic value (Hofer, 1972).

A series of convoluted bony plates, the turbinals and the nasal conchae, extend shelf-like into the nasal cavities, and function as support for the mucous membrane that lines the nasal cavity. Thus, enlargement of surface area in the nasal cavities is provided by these bony structures (conchae and turbinals). There are four pairs of ethmoturbinals in Tupaidae and Lemuridae, whereas Daubentoniidae alone possess five pairs. Ethmoturbinals lie one above another, with each successive lower pair located further backward; they are formed of cartilaginous replacement bone, as is the ethmoid. In all higher primates including *Homo* the number of ethmoturbinals is recuded to two pairs: These are positioned one above the other rather than one after the other. Above the ethmoturbinals lie the nasoturbinals. Nasoturbinals and ethmoturbinals are both parts of the ethmoid bone. Beneath these turbinals and extending from the lateral walls of the nasal cavities is another pair of turbinals. Even though this pair is a separate membrane bone, it is called the maxilloturbinal bone. The maxilloturbinal is homologous with the interior nasal conchae of human anatomy. The superior and middle nasal conchae in human anatomy are called the ethmoturbinals. As already mentioned, the surfaces of the bony structures within the nasal cavity are covered with a ciliated mucous membrane. This membrane functions in the upper part of the nasal cavity as the olfactory organ and is equipped with olfactory receptors. The lower portion of the nasal region functions mainly as a warm humidifier and cleaner for the air entering the respiratory system and is devoid of olfactory receptors.

In long-snouted prosimian primates the nasal cavity and the inner surface within the nasopharyngeal tract are comparatively larger than in higher primates. The difference is also expressed in the reduction of turbinals in the nasal cavity of anthropoid primates. However, the relative size of nasal cavities cannot necessarily be correlated with corresponding differences in smelling ability. It is not the size of nasal membrane surface that determines acuity of the sense of smell but the number and capacity of olfactory receptors per unit area of the membrane. Because knowledge of the relative olfactory abilities of primates is still restricted, the idea that large-snouted primates (prosimians) have better olfactory abilities than short-snouted primates (higher primates) has not really been proven, and there are other reasons for long snouts. In baboons, for example, the snout is considerably elongated in apparent correlation with enlargement of chewing apparatus and elongation of the tooth row and not with enhancing olfaction or respiration.

The ethmoid is a very fragile bone that is paired in early ontogeny but later fused into one single element. This bone not only forms part of the nasal cavity but also makes up part of the inner wall of the orbit and part of the anterior cranial cavity, the cribriform plate. This plate is perforated by many foramina that are penetrated by olfactory nerves. This sievelike arrangement gives the ethmoid its name, for *ethmos* is the Greek work for "sieve." It is joined posteriorly by the sphenoid, *sphenos* being the Greek word for a wedge. The sphenoid is a single bone in adult primates but consists of eight parts (two basisphenoid "Anlagen," two orbitosphenoids, two alisphenoids, two pterygoids) in early ontogeny. It is made up almost entirely of cartilaginous replacement bone and is a part of the inner and outer skull base as well. The sphenoid bone forms part of the hind wall of the orbits and the outside of the cranium. Within the medial body of the sphenoid, a groove is formed that is situated approximately in the center of the cranial cavity. This groove is part of the inner floor of the cranium and holds the hypophysis, or pituitary, of the brain. The sphenoid is posteriorly joined by the occipital bone, which surrounds the foramen magnum and which makes up the medial and hind parts of the skull base. Its front part joins the sphenoid, and the closure between these two bones is used by some as a feature for determining relative age, whether adult or juvenile, in primate skulls. Early in the ontogeny the occipital bone starts to form in five different centers: a central part, basioccipital; two lateral parts with the occipital condyles, exoccipitals; and a flat back part, the *"squama,"* which forms the back of the cranium. The *squama occipitalis*—or occipital bone— also appears in two parts very early in ontogeny, the lower part arising with the basi- and exoccipitals and cartilaginous replacement bone and the upper portion of the flat part develops as membrane bone. In contrast to this, all the other flat bones that form the cranial vault proper arise as membrane bones only. The pair of frontal bones does not fuse together along the midline between them (metopic suture) in many prosimians, but normally fuses in higher primates before they reach adulthood. A pair of parietals comprises the central upper part of the skull cap or vault. These articulate anteriorly with the frontals and posteriorly with the occipital. In those primates with well-developed skull musculature and small brain cases, we find bony skull crests. These crests enlarge the area for chewing muscle attachment. Bony crests are most developed in adult males, but the size and shape of these structures also vary individually. The temporal bones participate laterally with their flat parts—the squamae—in forming the sides of the cranial vault. These *squamae temporalis* develop as membrane bones. The basal parts of the temporal bones containing the ear region are of chondrocranial origin. As part of the skull base, the temporal bones are locked in between sphenoid and occipital. Three parts are combined to form the basal portion of the temporal bones. The petrosals are paired bones situated on either side on a line crossing the approximate center of the skull base. The petrosals contain the inner ear with the opening of the internal acoustic meatus. The middle ear bones (*malleus, incus,* and *stapes,* all cartilaginous replacement bones) are situated in this part of the temporal. The noun "petrosal" is derived from the Latin word *petrus* meaning "rock," a name justified in man and many higher primates because it is composed of dense rocklike bone. In prosimians and many of the New World monkeys, however, this part of the temporal bone is blown up like a balloon and called an auditory "bulla." Nothing much is known about the origin of these structural differences. Thus, we do not know whether there are functional differences between these structurally different morphological features. This also holds true for differences in relative size and stucture of the three ear ossicles. Thus, attempts to use *malleus, incus,* and *stapes* as taxonomic characteristics (Masali, 1968) must be regarded of uncertain value.

In Tupaiidae the "bulla" is formed by a separate ossification center, the so-called entotympanic bone (Spatz, 1964). Some have regarded this element as a part of the petrosal, but recent work by Cartmill makes it a wholly different bone. Thus, the bony floor of the auditory region is formed by the petrosal in prosimians and anthropoideans but not in tupaiids. Another part of the petrosal is the styloid process (cartilage replacement bone) that is especially well developed in modern man. The styloid process and the (unpaired) hyoid bone (cartilage replacement bone) are phylogenetically derived from visceral arches. The rear portion of the basal part of the temporal bone is known as the mastoid. It is developed as a very prominent and rounded process in man. Sometimes it also shows a certain prominence in mature specimens of genera *Papio, Macaca, Gorilla, Pan,* and *Pongo* but much less so than in man. The tympanic (ectotympanic) position of the temporal bone provides the bony ring for the tympanic membrane. The position of this ring (and membrane) is different in different primates and is also used in determining taxonomic affinities. Its size and shape have been regarded as useful characteristics in distinguishing different groups of primates. This region has recently been reevaluated by MacPhee (1977) and once again has become the topic of some controversy (Conroy and Wible, 1978). The tympanic ring is suspended within the bulla in Tupaiidae and Lemuridae.

In Lorisidae and Platyrrhini the ring is situated at the lateral edge of the skull and forms the margin of the bony auditory region (except in *Nycticebus* and *Loris*). In all Old World monkeys, apes, and man the ring is positioned rather medially, compared to the prosimians, and is extended outward into a bony external auditory meatus that has its opening at the lateral margin of the skull. A short bony meatus lateral to the bulla is also found in the Asiatic lorisid genera *Loris* and *Nycticebus*. In *Tarsius*, with its rather globular skull and comparatively centrally positioned bullae, one finds relatively long bony auditory meatus. In those primates that have a bony external auditory meatus when adult, the meatus forms only during postnatal life and the ontogenetically primary situation is as in platyrrhines. From this, we can see that it is not only the final form of growth in a bony structure that is useful in determining taxonomic affinities. Often, early developmental stages more clearly reflect the derivation of the structure concerned. On the temporal bone (*pars squamosa*) lies the facet for the articulation of the mandible with the skull, the glenoid cavity. This articular surface is located in front of the external auditory opening on the lower aspect of the posterior base or temporal portion of the jugal arch. A forward-directed extension of the jugal arch from the squamosal runs under that part of the actual jugal bone that extends back from the cheek. A downward-projecting bony process—the postglenoid process—is positioned in front of the external auditory opening and behind the articular facet (mandibular fossa). This postglenoid process makes up a rim or stop to the articular joint in all primates except *Daubentonia*, where this abutment is totally missing. Absence of this process in *Daubentonia* no doubt results from the functional relationships of the peculiar, rodentlike jaw and dentition of this primate.

Among primates, the mandibular—or glenoid—fossae are typically wide and flat, allowing the jaws to move through a three-dimensional range. The external margin of the orbits—or orbitae—is formed by the maxillary bone (below) the jugal (laterally) and the frontal (above). As we have discussed, among prosimians the orbits are not closed in behind except in *Tarsius*. The jugal and frontal contribute to a postorbital bar, and the lateral postorbital wall of higher primates and in *Tarsius* is also formed by these two bony elements plus the alisphenoid. The medial lower part of the orbital margin is formed by a paired membrane bone, the lacrimal—a small bony element that also forms the outer opening of the lacrimal canal. The canal—or "foramen lacrimale"— is normally positioned outside the orbits in Lemuridae. With the latter animals the lacrimal bone also extends beyond the orbital margin into the cheek area (Figure 4–3). In higher primates the lacrimal bone usually does not extend beyond the orbital margin or does so only slightly. Also, in Anthropoidea, the opening of the lacrimal canal is positioned within the orbit. Only in some individuals of the genera *Cebus*, *Macaca*, and *Papio* is the opening of the lacrimal canal frequently positioned within the margin or slightly outside the orbit. The roof of the orbit is formed by the frontal bone in all primates. The ethmoid is also present in differing degrees on the inner wall of the orbit in various primates. In Lorisidae, Tarsiidea, and higher primates as well as a few Madagascar lemurs (*Microcebus*, *Lepilemur*), the paper-thin lamella of the ethmoid (lamina papyracea) is positioned between the frontal and the maxillary bone. This place is occupied by an extension of the palatine in Tupaiidae and most Lemuridae and Indriidae. In Lorisidae, Tarsiidae, and higher primates the palatine participates with a small plate on the medial wall of the orbit behind the ethmoid. The palatine also forms only a small area of the hind part of the orbital wall in insectivores,

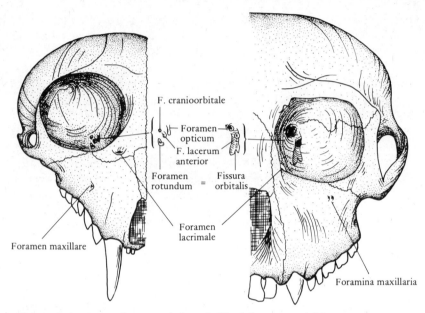

Figure 4–3 Frontal view of the skull of *Lemur* and *Macaca*.

where the frontal usually directly contacts the maxilla. These sutural combinations within the primates are, however, variable intraspecifically. Some anatomists, when sorting primates into groups, emphasize differences in the position of one basicranial foramen. This is the carotid foramen for the arteria carotis interna, a large artery of the head. The foramen's position varies in apparent correlation to changes in the shape of the petrosal (or bulla) in primates. Hence, the different positions of this foramen are regarded as of taxonomic value in recent and fossil forms. Sometimes, too much significance can be ascribed to positional differences of skull features. Many "characteristic" morphological details of the primate skull, such as the position of this foramen caroticum and the shape of the bones of the ear region (bulla, petrosal, and tympanic), are mainly dependent on the total shape of the skull, the position of the skull on the neck (foramen magnum), and the overall shape and size of the head. The already mentioned carotid artery runs up the neck and enters the skull near the ear region or bulla. However, the artery does not serve the ear region but the brain. Generally speaking, with the change from prosimians to higher primates, this foramen moves from the back and behind the ear region toward the center of the skull and in front of the ear region. Also the diameter of the artery and the foramen increases with the increase of relative brain size among primates.

Except in Lorisidae the artery runs through a bony tube penetrating the inner ear area or petrosal adjacent to the inner ear.

In Tupaidae, Indriidae, and Lemuridae—except Cheirogaleinae—the foramen caroticum penetrates into the skull behind the bulla. Thus, the foramen is called foramen caroticum posterius in these cases. *Posterius* in Latin means "behind." The bony tube that houses this artery splits up after penetrating the region of the ear drum. One of the two branches, the arteria stapedia, runs through the stirrup bone or stapes and then continues into the brain region (this penetration of the earbone gives it the name of stapedial artery). The second short branch after the split is called arteria promontorii (entocarotis or carotis interna) and is an artery of small diameter. It fuses with the arterial circle (Circle of Willis). In *Tarsius* and all higher primates the arteria stapedia is present early in fetal life but disappears later in ontogeny (Figure 4-4).

A stapedial artery is typical of primitive mammals, where it has branches in the orbital region and the upper and lower jaws. For Lorisidae and Cheirogaleniae there are two different interpretations concerning the internal and stapedial arteries.

One interpretation is that the internal carotid splits up into a stapedial and promontory artery before it enters the skull base, the arteria promontorii being of larger diameter than the stapedial branch (just the reverse of the relative proportions in Lemuridae). After the split, the promontory artery, according to this interpretation, enters the skull through the so-called foramen lacerum (a foramen that is absent in Lemuridae) that lies in front of the bulla, and the small stapedial artery enters through a tiny foramen behind the bulla (Werner, 1960).

Contrary to that interpretation, Saban (1963) maintained that the situation of the internal carotid of Cheirogaleinae and Lorisidae does not really differ from the arrangement of these arteries in Lemuridae. His description is that the internal carotid artery also enters the skull behind the bulla in Lorisidae and Cheirogaleinae. In Saban's view the artery entering the skull in front of the bulla is an additional branch off the stem artery "arteris communis." This branch is the so-called anterior carotid artery that, according to Cartmill (1975), is actually the ascending pharyngeal artery. This artery enters the skull through the foramen lacerum medium. The artery is of comparatively large diameter. In Tarsiidae the internal carotid enters into the center of the large bulla. The carotid passes into the skull relatively posterior of the petrosal in South American monkeys, whereas it enters the skull of Cercopithecoidea approximately in the middle of the petrosal and the skull of Hominoidea in front of the petrosal area.

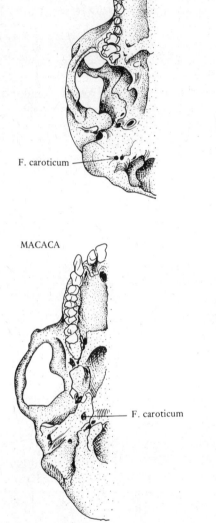
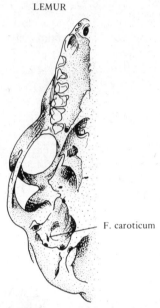

Figure 4–4 Base of skull with foramina of *Perodicticus potto*, *Lemur* and *Macaca*. Arrows indicate foramen caroticum.

SKULL MORPHOLOGY

The skull of Tupaiidae is generally narrow and elongated with the snout usually comparatively long and pointed. The foramen magnum is positioned at the back of the skull and opens in a backward direction. The eyes are surrounded by the postorbital ring only. The mandible is low and has a backwardly pointing angle.

Prosimii

Generally, in Lemuridae the skull is fairly elongated with a long tapering snout (foxlike), shallow mandibular rami, and complete post-orbital bars. Relative to monkeys, apes, and humans, the brain case is comparatively small and positioned behind the facial region. With these animals the orbits are situated somewhat back from the facial skeleton, or snout, and may be directed somewhat laterally. Thus, many lemurs have skulls that are narrow and elongated. The foramen magnum at the back of the skull opens backward. Also, the inner morphology of the lemur skull is elongated in correlation with the entire shape: the inner cranial floor remains almost straight (Figure 4–5). This means that the anterior part of the skull base (which is mainly formed by the ethmoid) and the posterior portion (consisting of the sphenoid and basioccipital) do not form an angle.

Overall skull shape among Indriidae is strikingly different from that of the Lemuridae. In indriids the skull is broader comparative to its length than the lemur skull, but the facial skeleton is still positioned directly in front of the brain case. Also, the mandible is much deeper and more robust in Indriidae than in Lemuridae, and when seen from above rather V-shaped, not U-shaped as, for example, among higher primates.

This condition, however, applies only to the shape of the mandibular arcade and not as clearly to the maxillary arrangement of the teeth, which overlap outward over the lowers and thus are set in a nearly U-shaped outline. There is no real occlusion of the anterior dentition in most prosimians, for the upper incisors are more or less reduced and the lower incisors in most cases highly specialized.

The skull morphology of the aye-aye—family Daubentoniidae—is highly specialized. This specialization is caused mainly by the relatively large brain and by the reduction of number and complexity of

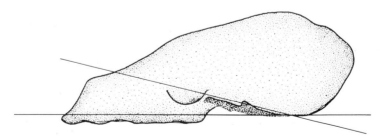

Figure 4–5 Outline of the skull of a *Lemur* showing the straight inner skullbase (stippled).

teeth in *Daubentonia*. As already mentioned, the anterior teeth of the aye-aye are unique among living primates: they resemble the rodent type of dentition. In accordance with this rodentlike adaptation, the entire skull of *Daubentonia* is atypical for a prosimian. For this reason the aye-aye was originally classified as rodent and not as primate (Cuvier and Geoffroy, 1795).

A high and transversely narrow snout characterizes the skull of *Daubentonia*. The face is also relatively short compared with the snout in Lemur skulls, and the brain case is globular and rather short. The relatively highly convoluted brain surface is impressed on the inside of the brain case. As in Lorisidae the foramen magnum of *Daubentoniidae* opens more downward than backward.

The skulls of Lorisidae, compared to the skulls of other prosimians, give the impression of being dorsoventrally flat. The interorbital distances are smaller in Lorisidae than in the Malagasy Lemuridae or Indriidae. This narrowness is also seen among lorisids in the lessened postorbital breadth of the skull—or "postorbital constriction." Moreover, among lorisids the snout does not taper toward the front as much as in Lemuridae and thus gives the impression of being less long and pointed than in the latter. Among Lorisidae the nasal bones are flatter than in Lemuridae. A characteristic elongation of the snout beyond the front end of the tooth row is found in species of the two lorisid genera *Arctocebus* and *Loris*. This phenomenon is brought about as a result of their comparatively large premaxillae, whose upper margins project forward. The nasals also enter this projection, thus forming a pipelike nasal opening. Additionally, the snout is rather narrow in *Arctocebus* and *Loris*. In *Nycticebus* the occipital is flattened and faces backward. Thus, of all Lorisidae the foramen magnum opens most directly backward in *Nycticebus*. Skulls of galagids resemble those of lorisids but slight differences can be detected. For example, with galagids the cranial vault is slightly more rounded than in Lorisidae, with the interorbital distance somewhat broader.

In *Tarsius* the shape of the skull is dominated by the huge eyes (Figure 4–6), which influence the morphology of its entire skull, just as does the peculiar dentition in *Daubentonia*. In fact, the volume of one eyeball of *Tarsius bancanus* is nearly as big as the volume of the entire brain of these animals (Sprankel, 1965). His measurements are as follows:

	Volume (cm^3)
brain without dura	2.14
eyeball I	2.03
eyeball II	1.81

Skull 127

Figure 4-6 *Tarsius bancanus* head, notice size of eyes, position of the (very mobile) ears and naked skin of the nasal area that is not extended onto the upper lip. [Photo courtesy of Heinrich Sprankel.]

It appears that the tarsier skull is the most globular skull of all prosimians, for the brain case is almost spherical and a short, small, and narrow snout adds to the globular appearance of the skull as a whole. As already mentioned, the huge and flaring eye sockets are nearly completely closed in *Tarsius*. This condition is the one exception to the rule that prosimians have postorbital rings only. Perhaps in correlation with its upright, clinging, and leaping habits, the foramen magnum of *Tarsius* is centrally located on the skull base. The bullae are fairly large and are located near each other and close to the foramen.

Anthropoidea

If one compares skulls of New World monkeys with those of Old World monkeys, one can see that the former usually have comparatively smaller facial skeletons than do Old World monkeys. Even so, we have one exception to this rule in species of the genus *Alouatta*, the howler monkey (Figure 4-7). In this genus the skull morphology is influenced by an enlargement of the hyoid bones of the throat. The lower jaw is unusually deep and the mandibular symphysis

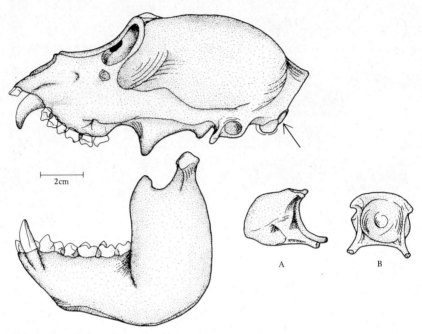

Figure 4-7 *Alouatta*, skull of a male together with (A) its inflated hyoid seen from left side and (B) hyoid seen from rear.

steeply oblique. The skull is flattened, and the tooth rows are tilted upward at the front, a condition that has been termed "airorrhynchy." In addition, the foramen magnum opens far more backwards than in other South American monkeys. Unlike other platyrrhine monkeys the posterior portion of the skull (occipital) behind the mandibular ascending ramus is shorter in howler monkeys than in other monkeys. The enlargement of the hyoid bone is greater in male howlers.

The petrosals of some of the South American monkeys of small body size have comparatively inflated, large bullae. In the temporal fossa of South American monkeys, there is usually a suture between jugal and parietal bones, as is frequently the case also in colobine monkeys of the Old World (Vogel, 1968).

The genus *Saimiri*—the squirrel monkey—has a very peculiar skull. It is the smallest cebid monkey, and the relatively large brain dominates the skull morphology, whereas the facial skeleton is relatively small. Thus, the foramen magnum is shifted toward the center of the skull. Also, the occipital portion of the skull projects farther backward than in most other monkeys, and thus the foramen projects straight downward, so that in *Saimiri* the foramen magnum is even situated relatively farther toward the front of the skull than in man. In *Saimiri* the large size of the brain is related to its greater complexity and to the fact that *Saimiri* is the cebid monkey with the smallest body size. The brain in Cebidae has a more complex structure than in the other South American monkey group, the Callitrichidae, which have a uniformly smaller body size.

Although they are big brained, the size of the masticatory apparatus of these small monkeys is unusually small even in direct proportion to the larger cebids. Thus, the unique skull morphology of squirrel monkeys appears to be a compromise between a small facial skeleton and a relatively large brain case.

In man the central position of the foramen magnum can in part be explained by the unique bipedal and upright human locomotion. This is not so in *Saimiri* (Biegert, 1963). Here we have a perfect example of two similar morphological traits found in two different types of primates but which have arisen for very different reasons. Additionally, the likeness in the position of the foramen magnum in *Saimiri*, *Tarsius*, and *Homo sapiens* shows that the central position need not, as was once thought, be an indicator of upright walking nor even of vertical clinging. There is another unique feature of the *Saimiri* skull and it is that all squirrel monkeys have a hole in the bony intraorbital (ethmoid) wall. This hole is usually more than 1.5 cm in diameter and cannot be found in any other extant primate. As yet, there is no clear explanation available for this structure, but the hole's presence in a late Oligocene Ceboid fossil—*Dolichopithecus*—makes it of ancient origin. The hole certainly cannot be in correlation with eye size because squirrel monkeys are diurnal primates and do not have enlarged eyeballs. In *Aotus*, with greatly enlarged eyes, it does not occur. In correlation with this feature of *Saimiri*, the interorbital breadth is, of course, much reduced (Figure 4-8).

Comparative to the general appearance of the skulls, the facial part of the skull is typically larger in catarrhines than in platyrrhines. Among the catarrhines the snout is frequently long and prominent, a feature that is especially marked in those forms—like the baboons and great apes that are comparatively large. In the case of these primates, the prominent anterior part of the skull, called the snout or rostrum, is not enlarged in correlation with an expanded olfactory region and sense of smell as it is, for example, among the dog family. Rather, its size seems related to the large masticatory apparatus among Pongidae and perhaps in the case of cercopithecids related to both the greatly enlarged canines and

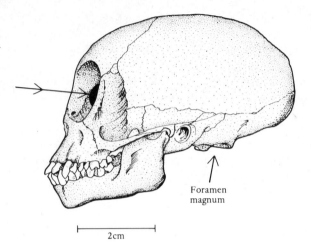

Figure 4–8 Skull of *Saimiri sciurea* with arrow indicating the large foramen between the orbitae.

associated chewing muscles in baboons (*Papio* and *Theropithecus*). Napier and Napier (1969) call the long snout in genera like *Tupaia*, the other tree shrews, and in Malagasy lemurs an "olfactory snout," and they call the long snout in baboons and Pongidae a "dental snout." Among the great apes the "dental snout" is relatively large when compared to brain size. This phenomenon is partly a result of their allometric increase in size relative to their ancestors. With increasing overall body bulk the masticatory apparatus also increases, but the proportionate size of the brain does not go up at the same rate, see, for instance, Biergert (1962).

Among Old World higher primates the petrosal is not expanded into a balloon like protrusion as in many New World monkeys and among Prosimii. Laterally, the petrosal of these animals is fused with the tympanic—or extotympanic—as a tube extending sideways to the external ear in all the Old World higher primates. It thus forms the external auditory meatus of the ear.

Among Anthropoidea the position of the lacrimal bone varies. In the cercopithecine monkeys the lacrimal bone entirely encloses the opening of the tear duct, or fossa lacrimalis, the drainage canal for the tear gland. A different arrangement of bones around the tear ducts occurs in both Colobinae and Hominoidea, where the maxillary and the lacrimal bones enclose the fossa equally on its two sides. In species of *Macaca* and *Papio* the lacrimal bone usually extends outside the bony orbital margin, but in the other primates it remains within the orbital wall. Almost all prosimian primates contrast with Anthropoidea in that the lacrimal fossa is situated outside the orbital wall, and the lacrimal bone containing it extends out over the orbital rim onto the face.

Skull

Behind the orbital region lies the temporal fossa, where the relationships of contact between adjacent bones varies among Old World monkeys. Here, in Cercopithecinae, one usually finds that the frontal bone meets the temporal bone in a suture common to both (Figure 4–9). In contrast, among Colobines the most common arrangement of these bones separates the two bones entirely for the jugal and parietal, themselves separated by a suture extending between frontal and temporal bones (Figure 4–9). Such contacts of different bones and the consequent sutures between them are typical for particular groups but can vary between individuals.

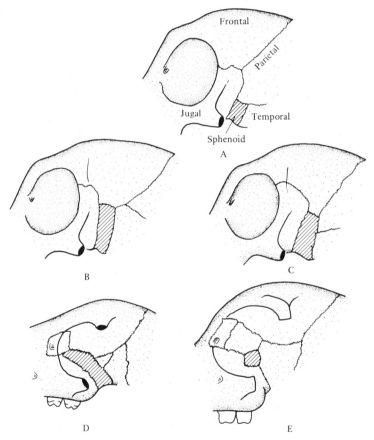

Figure 4–9 Variations of sutures in the temporal groove among Cercopithecinae (ali-sphenoid-striped): (A) jugal-parietal contact, (B) frontal-temporal contact, and (C) sphenoidal-parietal contact. Participation of the palate within the composition of the inner (medial) orbital wall (palate-stippled) as typical among: (D) Tupaiiformes and Lemurformes and (E) Lorisiformes, Platyrrhini and Catarrhini.

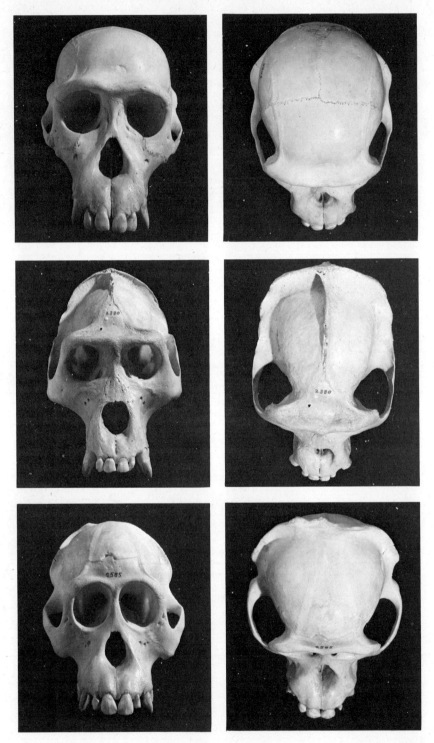

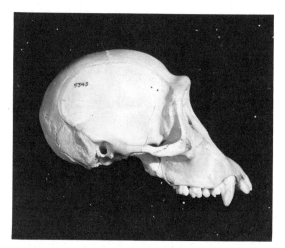

Figure 4–10 Comparison of skulls of male *Pan*, *Gorilla* and *Pongo:* frontal view; top view; and side view. (Skulls reduced to approximately same size.)

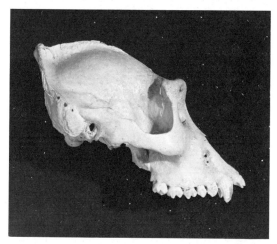

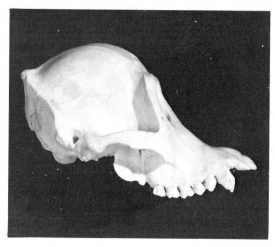

Lesser and Great Apes

At first glance one can see that the facial part of the skull, of lesser apes is comparatively small, and the whole skull is dominated by the large brain case. Also, the orbital openings of apes are comparatively large and typically larger than in all Cercopithecoidae. In the skull of the lesser apes at both sides of the nasal aperture, there are strongly developed vertical ridges. These ridges are formed by the roots of the large canines that are of equal size in both sexes of these lesser apes. Behind these ridges the maxillary bone is deeply depressed inward to form the fossa canina (canine fossa or groove). The mandible is comparatively shallow and slender in lesser apes, and the ascending branches or rami of those mandibles do not stand very high. In species of both *Hylobates* and *Symphalangus* the supraorbital tori—or eyebrow ridges—remain comparatively small.

In species of the two African genera of great apes, *Pan* and *Gorilla,* the orbits are also comparatively large and overhung by large supraorbital tori. Beyond this bony structure (torus supraorbitalis) that bridges across the forehead just above the orbital openings the skull is usually narrowed or constricted (see Figure 4–10). Orbital size is comparatively small in genus *Pongo* (the orangutan), and usually the brow ridges are either missing or weakly developed. In contrast to the chimpanzees (in which genus the two sexes are equal in size or very close to each other) there are considerable morphological and size disparities between the two sexes in the skulls of the gorilla and orangutan. Most of these differences are, however, related to the effects of the differences in body size between males and females. In adult males of *Pongo pygmaeus* and *Gorilla gorilla,* the skulls have distinctive sagittal and occipital bone crests. The height of the sagittal skull crest is correlated to the size of the temporal muscle (chewing musculature), and the extension of the occipital crest enlarges the insertion area for the neck musculature of these large animals. The size of both bony crests has an allometrically positive correlation to their body size. Such crests, for example, are not found among primates of small body size with relatively large brains: such small monkeys as *Saimiri,* the squirrel monkey, *Cercopithecus talapoin,* the swamp monkey, or among the lesser apes constituting the species of *Hylobates* and *Symphalangus.* In all such animals the braincase is large, and the area of the flat bones that shape the brain case have sufficient surfaces area for the masticatory musculature to insert itself without building up at the midline. A whole series of studies have shown that skull crests in primates are caused by topographical and functional requirements, see for instance Vogel (1962), Hofer (1965). Sometimes, but very rarely, cranial crests are found in exceptionally robust female gorillas and orangutans. The crests have even been reported a few times in chimpanzee skulls, and the crests can also occasionally be detected in the skulls of monkeys, especially the genera *Cebus* or *Colobus* (see Hofer, 1965).

chapter 5

The Brain

Primates have generally remained unspecialized in their postcranial organization. This is particularly so if trends within this order are compared with those of other groups of mammals such as, for example, ungulates. Thus, it seems that primates are the product of a combination of unspecialized morphology of the postcranial body with an increasingly specialized brain. There can be no doubt that one representative of the order primates (namely *Homo sapiens,* our own species) is distinct from other mammalian groups because of its unique progressive development of the brain.

Very different levels of brain construction are found among primates, varying from the simplest prosimian type of organization through that of monkeys and apes to a culmination in the great complexity to be seen in the brain of modern humans. The following trends generally characterize the primate brain: increase of size of the neencephalon (Greek for "new brain," or neocortex, the phylogenetically new and young parts of the brain cortex); increasing dominance of the optical system, presumably as a prosimian adaptation to an arboreal environment. Moreover, the visual system has become highly developed, attaining the ability to evaluate accurately the dimensions of space (three-dimensional vision). Increasing sensitivity of the tactile pads on hands and feet and, in some South American monkeys, the motor and sensory augmentation of a fifth extremity—the prehensile tail. Relative decrease of olfaction and taste. This decline in the senses of taste and smell has been often taken as almost a platitude for order primates. However, recent careful studies of taste abilities have been undertaken by D. Glaser of Zurich (1972 a & b). His studies make it possible to compare the sense of taste of many primates with those of nonprimates. From his new evidence it seems that statements about increasing reduction of olfaction and taste within the more advanced primates have been uncritically accepted and perpetuated. Such conclusions have been based largely on the macroscopic observation that in order primates, as one approaches man taxnomically, the number of nasal turbinalia and conchae is progressively reduced and that the correlated parts of the brain also become reduced in size. This apparent macroscopic reduction has been equated with an assumed concomitant reduction of smell and taste abilities, but such declines are certainly not true for the sense of taste.

Within the order primates we can recognize trends toward increase of the relative and the absolute volume of the brain. But when considering brain volume, we have to keep in mind that brain size is correlated to body size in a negative allometry: even as early as the eighteenth century it was understood that smaller individuals of the same species have relatively bigger brains than larger individuals.

It should also be pointed out that brain volume is only a crude measure of the evolutionary level of intelligence and achievement of an animal or a species. To regard brain volume as an indicator of cerebralization or, to word this differently, as a measure of the degree of evolutionary development of brains—even within quite specific groups of animals—is an erroneous or at the most, a superficial measure (Gould, 1975; Holloway, 1966; Starck, 1965). The usefulness and the limitations of cranial capacity estimates have been pointed out repeatedly by students of brain evolution (Edinger, 1961; Holloway, 1972; Starck, 1965; Hofer, 1972; and Radinsky, 1972; to name only a few). Even if brains of different but related genera are equal-sized and look similar macroscopically, they can differ considerably in the architecture of their internal structures. Brains are highly complicated organs, composed of parts that are different, and develop at different growth rates. Only the cyto-architecture (histology or cell and tissue structure) and electrophysiological mapping of brains can tell us about function of particular parts and consequently about degree of functional ability and development (Starck, 1965). Macroscopic comparison of brains can only give valid information about relative degrees of neurological development within taxonomically closely related groups.

Body size/brain size ratios are surprisingly similar in closely related groups of mammals. Stephan (1967, 1972) contrasted functionally different parts of primate brains to the equivalent areas among those of basal insectivores. He took for his "basal" group only such insectivores as are comparatively unspecialized with regard to brains. Thus, he only used animals such as the shrew—*Sorex*—and the hedgehog—*Erinaceus*—that have brains that can be regarded as useful basic models for comparison with primate brains. The basal terrestrial insectivores have served him as models for the definition of specializations in primate brains. Since 1960, Stephan and his collaborators have painstakingly measured and compared brain structures that are of evolutionary interest and whose functions are fairly well known. Contrasting the primates—except man—with other groups of mammals, Stephan (1972) concluded that a high degree of encephalization—that is, total brain weight—is not an exclusive characteristic of nonhuman primates because it is exceeded by that of some of the Odontoceti (toothed whales). Many other kinds of mammals have a stage of encephalization that is equal to that of prosimians. However, the trend toward high elaboration and development of the central nervous system, especially the cerebral cortex and its end-organs, is a striking characteristic common to all primate groups.

We have seen that comparison of brain volume has to be restricted to taxonomically closely related groups that are adaptively similar. If closely related genera show divergent adaptations, the results may be distorted. Thus, for example, two prosimians of small body size, *Tarsius* and *Microcebus*, would provide distorted results if they were compared indiscriminately in reference to their relative neocortex size. *Tarsius* would rank much higher in such a comparison than would *Microcebus*. This result arises not from an overall higher development of the neocortex of *Tarsius* but by the fact that the optical system is extremely enlarged in this form in accordance with its nocturnal feeding habits. It also has to be kept in mind that the "brain" cannot really be regarded as a single evolutionary or functionally homogenous entity similar to other organs such as, for example, liver or spleen. The brain is differentiated into parts that have the functional value of partially separate organs of varied structure and function. These different parts of the macroscopic entity "the brain" can evolve at quite different rates and independently of each other. Nevertheless, we know that these functionally different structures of the brain are often intricately interwoven.

The brain is enveloped by the brain case of the skull. Five major parts can be distinguished in the brain's gross morphology. These five subdivisions develop from three vescicles in the very early *embryo* (Greek for "offspring"): the forebrain or prosencephalon (*kephale* is Greek for "head," *pros* "front end," and *enkefalos* for "brain"), the mesencephalon or midbrain, and the rhombencephalon or hindbrain. These three vescicles soon give rise to two additional buds: the prosencephalon subdivides into the telencephalon and the diencephalon, and the rhombencephalon proliferates into metencephalon and myelencephalon. These five vescicles of the very early ontogeny correspond to the five major subdivisions of the adult mammal brain (compare Figure 5-1). The cavities of these vescicles of the early brain buds develop into the ventricles (fluid filled spaces) within the adult brain.

The most caudal portion of the hindbrain, the myelencephalon (*myelos* is Greek for "marrow" or "brain"), is also called medulla oblongata. The medulla connects with the spinal cord. The cranial portion of the hindbrain—the metencephalon—is subdivided into two distinguishable parts, the cerebellum and pons (Latin for "bridge"). Spinal nerves VII through XII originate from the medulla oblongata. Many sensory and motor tracts pass through this region connecting the higher centers of the brain. Cranial nerves V and VI originate from the area of the pons. Many of the nervous pathways function as projection tracts crossing over from one side to the other of the medulla. Also, many centers of the autonomic nervous system are located here, for example, the tenth cranial nerve—N. vagus (Latin for "prowl," "wander")—that wanders through the body to serve many of its major organs, reaching down as far as the colon.

The pons is positioned on the floor of the brain in front of the medulla and is attached to the overlying cerebellum by means of three nerve trunks on each side. The nuclei pontis (*nucleus* is Latin for "kernel") are major relay stations that transfer impulses from the cerebral cortex to the cerebellum.

The mesencephalon (*mesos* is Greek for "middle") or midbrain is situated between the diencephalon and the pons, and the only parts of it that are visible from the base of the brain in higher primates are the cerebral peduncles. In the center of the midbrain, the cerebral aqueduct passes through the mesencephalon and connects with the third and fourth ventricles. Lateral to the peduncles, the fourth cranial nerve arises and between them the third cranial nerve. The peduncles consist mainly of motor fibers descending from the cerebrum. Ascending sensory fibers passing to the thalamus lie deep to the peduncles. On the dorsal surface of the

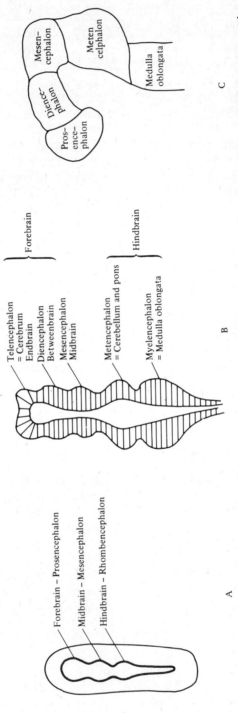

Figure 5–1 Early development of the brain: (A) three vescicle stage; (B) five vescicle stage; (C) five vescicle stage in side view.

midbrain, the corpora quadrigemina are located. The medial and lateral geniculate bodies that are part of the diencephalon (thalamus) are situated on the sides of the mesencephalon. In the diencephalon the substructure shows a sequence of differences in different extant primates. This can be understood in connection with the increasing development and elaboration of the visual sense. Functionally, the midbrain is a major relay station for auditory, visual, and tactile impulses.

The cerebellum is highly differentiated structurally: its surface is folded up into numerous deep fissures. Also, its lobes are considerably smaller than those of the cerebrum and are, therefore, called *folia* (Latin for "leaves"). The connections of the cerebellum with the cerebrum are numerous and close, as are the connections with the spinal cord. Major functions of the cerebellum are those of maintenance of muscle tone, maintenance of equilibrium, and positioning of the body in space as well as the coordination of body movements. Also, the cerebellar cortex is involved in coordination of visual, acoustical, and tactile sensations. The middle lobe of the cerebellum plays an important role in the coordination of the voluntary musculature and is connected by fiber tracts to the cortex of the cerebrum. The lateral hemispheres of the cerebellum are responsible for the autonomic regulation of equilibrium. The hemispheres are consequently joined up by tracts to the stato-acoustic organ of the ear and other parts of the brain.

Ontogenetically, the diencephalon, as we have seen, constitutes a part of the forebrain. The diencephalon is unpaired and consists of the thalamus, the epithalamus, the pineal body, the hypothalamus, and the infundibulum. The diencephalon also contains the third ventricle whose lateral walls are called thalami (thalamus). The thin roof of the third ventricle is the epithalamus, the floor of the third ventricle the hypothalamus. In fact, the entire thalamus is one of the most important sensory centers and relay stations of the entire body. The thalamus integrates tactile sensations, spatial perceptions, and feelings of pain and temperature as well as olfactory and visual functions. The thalamus also serves as a relay station for the motor control of facial and limbic gestures. A vascular structure—the so-called chorioid plexus—produces the cerebrospinal fluid within the epithalamus. A small cone—the pineal body—projects upward from the posterior part of the epithalamus.

Among other functions the hypothalamus contains autonomical centers of thermoregulation, circulatory functions, regulation of hunger feelings, digestion, and centers controlling sleep and wakefulness. The hypothalamus also carries out complicated functions governing hormonal secretion, arising in the pituitary stalk (infundibulum). The area of the hypothalamus contains higher centers of the autonomous nervous system, such as aspects of heat regulation, namely sweating and shivering. Control of the water, fat, and carbohydrate metabolism is carried out in cooperation with the posterior lobe of the pituitary, and sleep, sexual excitability, and emotions are also regulated in this area. The anterior part of the infundibulum or pituitary—also called the adenohypophysis is ontogenetically not a part of the hypothalamus of the brain. This anterior part of the hypophysis produces a number of hormones. Also, in the area of the diencephalon, we recognize the optic chiasma.

Most obvious are the evolutionary and ontogenetic changes in the cerebrum. The cortex (pallium) or mantle of this main part of the brain is smooth only among the smallest prosimians and the smallest monkeys. The construction of the pallium is as a superficial layer, and it does not become thicker than about 5mm. Its essential characteristics are those of a surface integration tissue. With increase of the number of surface cells, the number of afferent and efferent fibers increases also. If the number and density of such fibers gets too high to be accommodated within a certain area, then the spacing between neural cells has to be increased. In this case the pallium does in fact grow two-dimensionally, and if the space available within the brain case is insufficient for such expansion of the surface tissue, the pallium begins to fold up. By these means, an increase of brain volume is avoided and the surface character of the tissue maintained.

There are two reasons for increase of the number of neurons in the pallium of the brain. One: if the body mass increases, the volume of peripheral organs that have to be innervated, grows concomitantly. Two: an increase in neuron number can result from a higher elaboration of brain functions. Because enlargement of the cerebral cortex can be brought about by both these rather disparate factors, it is obvious that the reasons sought for brain volume increase and folding of the cerebral cortex have to be determined carefully.

The cerebrum itself consists of two hemispheres that are divided by a deep longitudinal fissure. The cerebrum is also separated from the cerebellum by a transverse fissure. At the bottom of the longitudinal fissure, the two hemispheres are connected with each other by several commisures and by a sheet of white cerebral tissue, the so-called corpus callosum. The rounded convex convolutions of the folded brain surface are called gyri, and the deep valleys between them are sulci. The gyri as well as the sulci, in different regions, have been given names. It is necessary to know a few of these in order to delimit macroscopically important cortical areas and their functions. The relative development of these areas helps us to understand evolutionary trends within primate brains.

When we look at the brain from the lateral aspect, a very prominent fissure begins just above the temporal pole and extends obliquely backwards and upward. This fissure is called the sylvian fissure; inside its posterior part is a hidden area called the insula. This fissure separates the temporal lobe from the major portion of the cerebral hemispheres in front and above it. Another major fissure that is not found in most prosimians is the so-called central sulcus or fissure of Rolando. The sulcus starts at the highest point of the hemispheres running down and slightly backwards to end above the sylvian fissure close to its midpoint. This fissure also separates the frontal from the parietal lobe of the brain. The gyrus in front of it is called the precentral gyrus and contains the motor region of the cerebral cortex. The gyrus behind the central sulcus, or postcentral gyrus, contains the area of tactile sensation.

The parieto-occipital sulcus separates the occipital lobe from the parietal lobe. Posteriorly on the medial or inside the surface of the hemispheres, a short sulcus is directed obliquely upward and forward from the lowest and hindmost pole of the hemispheres. The short sulcus meets the parieto-occipital sulcus at its base. The posterior walls of this so-called calcarine sulcus are occupied by the visual area of the cortex. Another prominent sulcus on the inner aspect of the hemispheres is the sulcus cinguli. It outlines the shape of the corpus callosum. The adjoining gyrus above this sulcus contains an area of smell association.

In structurally primitive mammals we find the neopallium to be smooth and without fissuration. This situation is also called lissencephaly (*lissos* is Greek for "smooth"). Fissuring and concomitant proliferation of lobes increases with the enlargement of the neopallium. The evolutionarily progressive cortex of the forebrain is called "neo" pallium because it is an evolutionarily young (that is, new) acquisition of the mammal brain. The living brain is positioned inside an enveloping cerebral fluid. In many Hominoidea the pattern of sulci and gyri is not imprinted into the endocranium as it is in lower primates. This has been thought to be the reason that fossil and extant endocranial casts of hominoid primates provide us with less information on surface structure than do those of lower primates (Radinsky, 1972; Starck, 1974). Nevertheless, this lack of detail in hominoid brain endocasts cannot be entirely explained and is not totally understood (Radinsky).

All comparisons in the following discussion of some of the

major differences that can be observed in the morphology of the brain among living primates are based on, and related to, a comparison with the brain of generalized insectivores. It has to be kept in mind, however, that no contemporary animal can be regarded as a good model for an ancestral form for any other contemporary animal, because both have had an equal amount of time for achieving present-day evolutionary development (Figure 5–2).

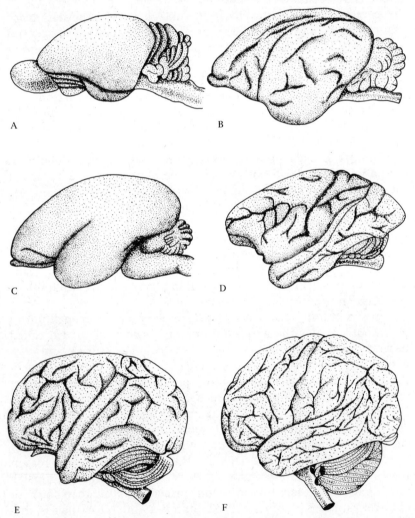

Figure 5–2 Diagrammatic comparison of brains (reduced to approximately the same length): (A)*Tupaia*; (B) *Lemur*; (C) *Callithrix*; (D) *Macaca*; (E) *Gorilla*; and (F) *Homo*.

TUPAIIDAE

Tupaia (Tree Shrew)

Relative to other insectivores in this group, there is a slight reduction of the olfactory region. The olfactory bulbs are situated frontal to the brain, and these and the olfactory tubercles are comparatively small. In tupaias the optical area is large in correlation with a well developed visual sense in these animals (Tigges, 1963). As a result of the comparatively large eyeballs, we find an "impressio orbitalis." Tupaias also show an enlargement of the neopallium in the temporal and occipital regions. However, the surface of the pallium is still lissencephalic, perhaps because of small body size. The cytological structure of the cortical-neural layer of the neopallium is more complex than in generalized insectivores. As we have already discussed, the optical area is rather large. However, frontal and parietal portions of the cortex remain small compared to those of higher primates. In the mesencephalon the colliculus superior of the tectum and the corpus geniculatum laterale are well developed. In fact, the organization of these rather highly advanced areas of the visual brain is more progressive in *Tupaia* than in many prosimians.

Ptilocercus (Pen-Tailed Tree Shrew)

Ptilocercus is neurologically less advanced than *Tupaia*, its close relative. The olfactory region is rather larger and the lateral geniculate body is smaller than in the latter.

LEMURIDAE AND LORISIDAE

Among extant prosimians the smallest representative, *Microcebus*, has what seems to be the simplest brain, probably because of its small body size. The mouse lemur's brain is lissencephalic. A deeply engraved sylvian fissure is present. The olfactory bulb is reduced and partly covered above by the frontal lobes of the brain.

In genus *Lemur* the olfactory region is more reduced than in *Microcebus*. The neopallium is convoluted and exhibits sulci that are mainly longitudinally directed. Many present-day prosimians retain well-developed olfactory regions (as, for example, *Dauben-*

tonia, Nycticebus, and *Galago*). Also, the long and snoutlike nasal region of *Lemur* and *Propithecus* is still structured like that of animals with a highly developed olfactory sense (Starck, 1962). Long snouts, however, are not necessarily correlated to olfactory acuity. Only the size of olfactory bulbs and extent of olfactory turbinalia with their receptors can give clues to the degree of olfactory sensitivity.

In *Daubentonia* the olfactory bulb is overlapped by the frontal part of the brain. This results from a highly specialized construction in the aye-aye skull. The olfactory region is not reduced, but the comparatively large olfactory bulbs are deflected downward, and thus, are covered by the frontal part of the pallium from above. The brain of the aye-aye is relatively larger and more convoluted than in any other prosimian, according to Stephan and Bauchot (1965). Among the non-Malagasy prosimians, *Nycticebus* and *Loris* have more cerebral convolutions than do the brains of galagos.

TARSIIDAE (TARSIERS)

In many features tarsiers are extremely specialized. Such specialization clearly appears in the configuration of their brain. Both skull and brain reflect that *Tarsius* is a highly visually oriented animal. As we have already noted, Starck (1953) and later Sprankel (1965) have shown that the combined volume of the orbitae, or eyeballs, greatly exceeds that of the brain.

Compared to other prosimians, the olfactory region of the tarsier brain is reduced. The pronounced separation of the temporal lobe is not brought about by a high degree of organizational development there, but mainly by an indentation made by the huge orbits in this region. There are no fissures on the surface of tarsier brains except the calcarine fissure that is located posteriorly on the medial side of the hemispheres. Half of the neopallium in tarsiers is occupied by the optical-cortical areas. The structure of the cortex is not more elaborate than that in the mouse lemur, *Microcebus.* The secondary areas in the temporal and parietal regions of the pallium are also small and not elaborate.

The foregoing shows that the extreme development of one of the special senses, in this case the optic, can result in considerable modification of the corresponding areas in the central nervous system. The remaining principal characteristics of the tarsier brain are not more advanced than those of other prosimians.

ANTHROPOIDEA (HIGHER PRIMATES)

South American marmosets—Callitrichidae—appear to have the most primitive brains of any higher primates. However, they do have an elaborate, expanded neopallium. "Reverse" allometry may be the cause of a primitive appearance to the brain, especially in its lack of folding. The olfactory regions of the brain are more reduced than in Old World prosimians. Areas of integrational processes in frontal, temporal, and parietal regions are enlarged, especially the temporal lobes. The cytological substructure of the cortical areas appear to be more complicated than those of prosimians. All members of Callitrichidae show a deep sylvian fissure and in addition a few other shallow fissures on the brain surface. In small marmosets only the calcarine fissure can be recognized. In the owl monkey—*Aotus*—a reflection of the nocturnal habits is expressed in a distinct expansion and elaboration of the optical areas.

The development of the brain in both the ceboids and the Old World monkeys appears to have reached a rather similar stage. Except for the smallest marmoscts of South America, both groups show a higher degree of fissuring and consequent gyral formation in the neopallium than do most prosimian groups. However, separation into clear-cut successive grades of organization is not possible for some prosimians, such as certain lemurids; the indriids and *Daubentonia* have more convoluted neocortices than do many ceboids. In contrast, the cerebrum has attained a higher degree of elaboration in ceboids than in prosimians; this condition is also true for parts of the cerebellum and the region of the thalamus. The lateral geniculate body shows high differentiation, apparently in correlation with the attainment of stereoscopic vision in these primates. With the ceboids the olfactory organ and the olfactory centers have decreased in size and importance. In these higher primates the tactile sense has become more elaborated than among prosimians. Whereas the major brain fissures in prosimians of large body size show predominant arrangement in longitudinal direction, their arrangement in monkeys tends to be more radial. It is most difficult to try to homologize the gyri and fissures of the brain in different mammals, except in those species where detailed cortical mapping has been done. The general agreement seems to be that fissures occur predominantly between areas of the brain cortex that undergo expansion. Among anthropoid primates the South American Callitrichidae have the most primitive brain. It is however, larger than that of prosimians of similar body size. Marmosets have about the same body weight as do some of the African bush babies, but the brain volume of marmosets is about three times as large as typifies African bush babies, see Starck (1965). The callitrichid neopallium is more expanded than that of prosimians. Their cerebellum and neencephalon are covered by the occipital lobe of the cerebrum, and the temporal lobe is large. With marmosets the olfactory bulbs and the paleocortex (the phylogenetically old part of the cortex) are much more reduced than, for instance, in *Tarsius*. Individuals of *Cebuella*—the smallest genus of the Callitrichidae—do not have any fissure patterns on the brain surface. Also, the cortex of *Cebuella* shows a higher differentiation in cell structure when compared to that of prosimian primates. Species of Callitrichidae of larger body size than the pygmy marmoset—*Cebuella pygmaea*—do have a well-defined sylvian fissure.

As we have noted in the South American owl monkey, *Aotus trivirgatus*, we find—in correlation with its nocturnal activity patterns—a highly developed optical system. In contrast, the olfactory region of the owl monkey has undergone even more reduction than with marmosets.

Starck (1965) stressed the similar stage of brain development seen in two of the main segments of Anthropoidea, the Old and New World monkeys (except Callitrichidae). Both groups represent a very similar evolutionary grade with respect to relative brain to body size, and in an elaboration of the visual systems exhibiting a high occurrence of parallelism. Even though these two groups are similar throughout in organizational level, the New World monkeys show a more varied differentiation into adaptive types than do the relatively uniform Old World monkeys. Among Platyrrhincs of family Cebidae, it is the subfamily Atelinae that exhibits the highest development of neural organization (Hofer, 1958), clearly more than can be found among the cebines, alouattines, and pithecines.

Among New World monkeys the Alouattinae and Atelinae are assisted in their locomotion by a fifth extremity: the tail. It can be shown that in both of these subfamilies, not only the musculature but the vascular system and the skeleton and nerve supply of the tail are all specialized for these prehensile faculties (see, for instance, Ankel, 1972; Wrobel, 1966; and Leutenegger, 1973). As early as 1907, C. Vogt and O. Vogt demonstrated that in *Ateles* the cortical motor representation area for the tail is larger than the motor area for hand and leg together. *Atels* is also the most skillful of the New World monkeys in tail manipulation. Structural evidence for this is provided, as well, by the presence of a short sulcus additional to all the others in the motor cortex of species of this genus, separating an expanded tail control area from that of the rest of the body.

In both Old and New World monkeys large integration and relay areas occur within the frontal, parietal, temporal, and occipital lobes. The conformity of many details in the brains of the two groups of monkeys is high. Their olfactory region appears to be reduced, and their optical areas highly developed in accordance with the elaboration of three-dimensional vision (the geniculate body shows a more refined structure than in prosimians and in marmosets). In addition to all the other similarities, the fissural pattern of the neopallium is very similar in both of these great groups of monkeys; in these groups especially, the larger cebids resemble Old World monkeys in their sulcal patterns.

One major trend within higher primates is the elaboration and enlargement of the frontal lobe. Concordant with this is the development of a more and more pronounced forehead that culminates in the greatly enlarged forehead of man (*Homo sapiens*). In Hominoidea and among some Colobinae, according to Starck (1974), the enlargement and differentiation of the parietal lobes is pronounced, more so than in other Old World monkeys. It appears that the sulci of the neopallium actually delineate areas of different function: for example, the central sulcus as the boundary between functional/motor area and somatic/sensory area.

The progressive complication and elaboration of the brain structure in Pongidae, possibly related to body size increase, is further developed in Hominidae. The prefrontal, the temporal, and the parietal lobes together have additional organization. The frontal lobes show the greatest expansion, even though it has not yet become possible to define in a clear-cut manner the functions of the frontal lobe. The agreement among most students of the brain and its functions seems to be that the frontal lobes of the neopallium govern those qualities in man that make him human. These qualities include perhaps exclusively human characteristics as foresight and insight, initiative, and advanced learning abilities. In addition, the motor speech center is also located within the frontal lobe. This lobe is seen most expanded in *Homo sapiens* (modern and fossil humans), the only animal with a really expanded forehead. As we have seen among some anthropoid apes, for example, the orangutan (and also with the spider monkeys), a slightly developed forehead occurs. But this slight forehead expression might be related to factors other than large frontal lobes such as short snouts, relatively large total brain size, or insertions of the muscles operating the jaw mechanism. Even in the more primitive extinct species of *Homo*, *H. erectus*, and *H. habilis*, forehead expansion is not really comparable to that of *H. sapiens*.

Right up to the present moment, there has been no certainty of finding correlations between morphological structures of the hominid brain and psychological performance such as memory or the ability to form abstract concepts. In Hominidae the morphologically primitive or generalized five-rayed hands have become the executing organs that have participated in the creation of culture and civilization at the direction of an advanced, higher specialized brain.

All living primates; prosimians, monkeys, apes, and modern men have gone through the same time span of evolutionary development and have brains that are well adapted to their specialized needs. Somehow, different or accelerated rates of change have led our own species to the forefront of the animal kingdom.

chapter 6
Postcranial Skeleton

VERTEBRAL COLUMN AND THORAX

The vertebral column is the axial or longitudinal support skeleton of the body of every vertebrate. This column links the head, the fore- and hindlimbs, and if present, the tail. The column is a combination of a series of numerous separate and morphologically different elements, the vertebrae. The vertebrae are articulated with each other. Thus, the "column" is both stable and flexible at the same time. Each vertebra is made up of different parts. The vertebral body, core, or centrum, is the largest part in most vertebrae, and it serves mainly weight-bearing and shock-absorbing purposes. The vertebral bodies are connected to each other by means of elastic cushion-like pads of fibrocartilage—the so-called intervertebral discs. These discs enable two vertebral bodies—next to each other but separated by the intervening intervertebral disc—to move slightly. Thus, both minor circular and slight to-and-fro movements are possible between adjoining vertebral bodies. At its top (pointing backward in bipedal man) the vertebral body carries a bony arch, also called a neural arch. This arch covers and protects the major postcranial nervous system, the spinal cord. On each side near the base of the neural arches, we find two bony articular processes pointing in the anterior direction and, at the back, two pointing posteriorly. These processes are called zygapophyses. These articular processes are positioned differently in different regions of the vertebral column.

Table 6-1 Average Number of Vertebrae

| | Insectivora | Prosimii | | | |
| | Tupaiiformes | Lemuriformes | Lorisiformes | | Tarsiiformes |
			Lorisidae	Galagidae	
Cervical	7	7	7	7	7
Thoracal	13	13	16	13	13
Lumbar	6	7	7	6	6
Sacral	3	3	7	3	3
Caudal	25	25 (Indri 10)	9	25	29

| | Anthropoidea | | |
	Callitrichidae	Cebidae	Cercopithecoidea
Cervical	7	7	7
Thoracal	13	14	13
Lumbar	7	5	7
Sacral	3	3	3
Caudal	27	30	17
	Hylobatidae	Pongidae	Homo
Cervical	7	7	7
Thoracal	13	13	12
Lumbar	5	4	5
Sacral	5	6	5
Caudal	3	3	4

Data from Schultz (1969).

Thus, each vertebra has three points to articulate with the vertebra in front of it and three articulating with the vertebra behind it. There are, however, some exceptions to this rule. We count the vertebrae within each vertebral column from cranial to caudal; the first one, connecting the head with the column—called the atlas—is the first exception from the already named rule and articulates differently with the second vertebra. There is no vertebral body in the first vertebra of the column; instead, this vertebra is ring shaped. Lacking the vertebral body, the first vertebra is also without an

Postcranial Skeleton

intervertebral disc. It does articulate, however, with three articulations on the second vertebra. There are two articular facets facing forward and embracing the condyles of the skull, and two articular facets directed backward joining corresponding facets of the second vertebra or axis. All four of these articulation facets are positioned within the ventral half of the first vertebra. The front end of the vertebral body of the second vertebra is elongated into a conical projection, called the *dens*, meaning "tooth" in Latin. This toothlike structure fits into a rounded opening on the lower and inner side of the first vertebra. A strong cross-ligament covers the dens from above, and thus separates it from the lumen of the neural arch of the first vertebra. At its hind end the second vertebra articulates with the adjoining third one in the manner already described as typical, with an intervertebral disc and a pad of fibrocartilage between the vertebral bodies and with two articular processes on the neural arch.

There are two other exceptions to the rule of vertebral articulation. In the pelvis, where the column is connected with the bony girdle for the hindlimbs, the vertebrae are commonly fused to each other. Thus, several vertebrae lose their articulations and are continuous or locked into each other, forming a solid and elongated element within the vertebral column, the sacrum. The third case of an exception in the manner of articulation between vertebrae is found within the tail. The peripheral tail vertebrae lose all the bony processes; these vertebrae are represented only by elongated vertebral bodies, connected to each other by intervertebral discs. This condition allows free mobility of the segments in all directions.

The spinal column is subdivided into five regions in primates, as in all tetrapods: the cervical or neck region, the thoracic or chest region, the lumbar region, the sacral region, and the caudal or tail region. All these regions show certain morphological characteristics.

Cervical region. The cervical region is a combination of seven vertebra in almost all mammals. Even this comparatively stable number of seven sometimes varies in ourselves and with other individual primates. The vertebral bodies of the cervical region are more or less square in cross-section and saddle shaped: In front they overlap the joining vertebral bodies from the sides, and in back the lower edge of the cervical vertebral body protrudes over the adjoining vertebra in a shingle-like manner.

The transverse processes of the cervicals (2–6) are pierced by cranio-caudal foramina: Foramina costotransversaria, through which the arteria vertebralis passes to the head. The points of these processes are double in vertebrae 4–6. The articular facets at the sides of the neural arch are positioned horizontally but are turned slightly upward and sideways. These facets are also spread far apart from each other: the neural arch diameter is widest in this region of the spine, and the articular facets do not lie above the vertebral bodies, as in all remaining regions, but lie higher up and arise on the arch at a point lateral to the sides of the centra. The articulation between the basioccipital of the skull and the first vertebra allows mainly antero-posterior nodding movements. The atlas-axis articulation is the place for rotary movements of the head. This articulation functions, for example, when humans shake the head in disagreement.

Thoracic region. The distally adjoining vertebral region to the cervicals is the thoracic region. It is characterized by the presence of ribs articulating with each thoracic vertebra. The vertebral bodies are typically heart shaped if seen from the proximal or distal end. The transverse processes of the thoracic vertebrae have at their ends articular facets for one of the articulations of the ribs: the so-called *tuberculum costae*. The head of each rib articulates with the side of the vertebral body; in the middle of the region, this articular facet is divided and runs across two adjoining vertebral bodies. Running down the series, the vertebral bodies gradually elongate craniocaudally, their diameters increasing as well. The degree and rate of size increase in the vertebral bodies vary between primate species. The neural arch decreases its diameter in some primates in a craniocaudal direction, as is the obvious case in humans.

Thoracic vertebrae are recognized by the presence of articular facets for the ribs on the vertebral bodies and by transverse processes. The paired articulations situated upon the dorsal arch are positioned above (dorsal to) the vertebral centra. Seen from the cranial or caudal aspect, these articular facets lie more or less flat above the dorsal plane of the vertebral body, with their inner side slightly above their outer side. The bending of the ribs shows distinctive differences between prosimians and monkeys on the one hand, and apes and man on the other. The ribs are much less bent near their articulation to the vertebra itself in the former than among the hominoids. The high degree of bending of the ribs (along the long axis) in apes and men results in a barrel-shaped thorax that is broad from side to side in these primates, in contrast to the transversely narrow thorax of prosimians and monkeys. Additionally, this difference in rib-bending brings about a different position of the vertebral column within the thorax. In prosimians and monkeys the ribs hang down from the vertebral column. In apes and men the vertebral column is positioned more or less within the thorax because the ribs ascend from their articulation in a backward direction. The shape of the thorax is also dependent on the length of the vertebrae. Also the length of the transverse processes influences the distance between the two articulations between ribs and thoracic vertebrae and thus the position of the rib itself. Incipient broadening of the rib-cage can also be found, apart from hominoids, in a few prosimians (*Tarsius* and *Perodicticus*) and a few monkeys (*Ateles* and *Colobus*). Exclusively in apes and men, however, the broad barrel-shaped thorax is also expanded in its ventral bony elements, the sternum or breast bones. Because of this broadening, hominoids have been called "the Latisternalia." In many primates the number of rib-bearing vertebrae is not identical with the number of vertebrae showing the typical thoracic articulations; the position of articulation changes within the rib-bearing area. The articular processes on the neural arch, unlike all the other bony processes of the vertebral column, change their position within those single vertebrae located at the margins of each vertebral type within the column. Thus, they maintain the same positions within a particular vertebral type.

The first vertebra in the thoracic region has frontal articular processes in the same position as do the cervical vertebrae; at its hinder end, the first vertebra has articular processes in the position characteristic of the thoracic region. The articular processes at the front end are positioned in one of the distal thoracic vertebrae, as is typical for thoracic vertebrae. In this same vertebra the articular processes at the back are angled steeply to the midsagittal plane, as is typical for all lumbar vertebrae; see Figure 6–1. It is of functional importance that the direction of these articular processes does not change gradually from vertebra to vertebra. Because the position of the articulations does not change gradually, the functional stability of an entire region is maintained. There would be no functional stability without this abrupt change in orientation of the articular processes.

Lumbar region In lumbar vertebrae the position of the articular processes is also steep, enclosing narrow angles with the median sagittal plane. Thus, in this region, there is strictly back and forth movement between adjoining vertebrae (Figure 6–2). The main characteristics of lumbar vertebrae are the laterally projecting transverse processes. Here also, dorsal spines are usually very robust and high.

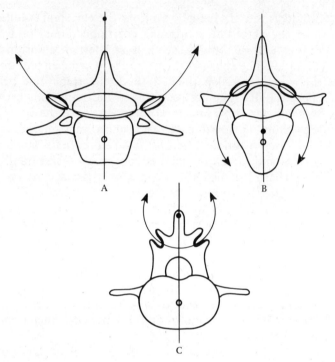

Figure 6–1 Diagram showing the different positions of the articular facets of the different regions of the vertebral column: (A) cervical region; (B) thoracic region; and (C) lumbar region.

Sacral region. The fourth region of all mammal—and thus primate—axial skeletons is the so-called sacrum. Here we have fusion of several vertebral segments into one elongated element. The sacrum is not only the longest element of the vertebral column, but also it connects the axial skeleton with the pelvic girdle and the hindlimbs. This is in fact the only solid connection of the vertebral column with the remainder of the postcranial skeleton. The transverse parts of the sacrum are enlarged into winglike bony protrusions that develop a close connection with the iliac blades. All articulations between neural arches within the sacrum are fused. Between adjacent pairs of sacral vertebra, an intervertebral foramen is located in the lateral transverse wing of each side. These foramina communicate with the neural arch and are open dorsally and ventrally. The neural spines are fused also; when not totally fused, they are at least connected at their bases. The neural arch articulations of the caudal region have the same position as the

Postcranial Skeleton 157

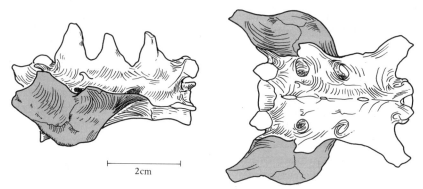

Figure 6-2 The sacrum of *Cercopithecus* (front and to the left, articular surface with the pelvis shaded).

articulation within the lumbar region. The neural articulations at the front and hind end of the sacrum, however, tend to be angled less steeply than those in the lumbar region. The same holds true for the position of the articular processes in the first caudal vertebrae, which, however, continue generally to have the lumbar position.

Caudal region. Tail or caudal vertebrae are typically smaller than all other vertebrae. Among primates this region is the most variable one in the number of segments that makes it up. Only a few of the first caudal vertebrae have a fully developed neural arch; the arch decreases rapidly farther down the tail vertebrae. There is evidence of a direct correlation between the total length and function of the primate tail (prehensile tails in some South American monkey species) and the number of first tail vertebrae that are roofed over by neural arches. These same tail vertebrae have transverse processes and dorsal spines. Tail vertebrae are characterized by so-called ventral arches, haemapophyses or chevron bones. These bony structures are short, V-shaped bone clasps (Figure 6-3). However, farther out toward the tip of the tail, these structures become two small separate bony nubs. Chevron bones

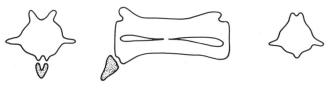

Figure 6-3 Proximal tail vertebra with chevron bone (gray).

appear exclusively in the caudal region. They are situated under the front end of the vertebral bodies, and they are often lost during the process of preparation of a skeleton, especially the more distally situated chevron bones, which are very small. Chevrons can, however, be seen easily in x-ray photographs. The chevron bones embrace the ventral arteries and nerves of the tail. The bodies of the first caudal vertebrae are comparatively short. Backwards, the bodies increase in length, whereas the bony structures (for example, the neural arch and its protrusions), the transverse processes, and the ventral arches are reduced in size. The increase in length of vertebral bodies goes to a vertebra of maximal length. This longest vertebra has different positions in different primate genera, and these positions are also determined by the length and function of the tail. Farther back from the longest vertebrae, the length and diameter of these segments decreases. As we shall see later, the last tail vertebrae of African monkeys are very thin and short but longer than those of prehensile tailed cebids.

Generally, the vertebral columns of primates are fairly uniform in their morphology. There are, however, some specializations.

Within the prosimians we find specialization in the Tarsiidae and Lorisinae. All the other living prosimians (Lemuridae, Indriidae, Daubentoniidae, Galaginae) have more or less uniform and undifferentiated vertebral columns. The uniformity of the vertebrae, ribs, and breastbones makes it impossible to easily identify the family, genus, or species of vertebral columns or—even less—of single vertebral elements.

In *Tarsius* the vertebral column is specialized but principally in the cervical region. There the articular processes between the hind end of the third cervical vertebra through the front end of the first thoracic vertebra (eighth vertebra in the row) are positioned differently than are those of all other primates. These articular surfaces are positioned in the same plane, with that of the ends of the vertebral bodies in *Tarsius*, that is, in a plane that forms a right angle with the median sagittal plane through the body. This position of the articulations is presumably partly involved in the ability of *Tarsius* and to turn its head about 180 degrees from the normal forward position, and look straight back over the shoulders without moving the body. This ability is superficially similar to the same movement in owls. The morphology of the vertebral column of *Tarsius* and owls, however, is totally different and thus is not structurally comparable. It has been observed in other primate genera (*Cebuella, Propithecus, Avahi*) that they are able to turn their heads backward almost as well as *Tarsius*, but none of them show this peculiar position of the articular processes of the cervicals. The thorax of *Tarsius* appears to be broader than in nonhominoid primates. This shape is not the result of heavily bent ribs, but it is produced by the relative length of the lower ribs and the unique length of the distance between *tuberculum* and *capitulum costae* (Figure 6–4). all presacral vertebrae are comparatively short and lack a distinct keel (in the middle of the underside) on the lumbar vertebrae.

In spite of the fact that *Tarsius* uses its tail as a support when clinging to an upright branch in a vertical position, there seems to be no difference in the morphology of the tail vertebrae when compared to other long-tailed primates. In cebids that use their tails as a fifth limb, the vertebral morphology shows changes adapted to function, as we shall see.

In their vertebral columns the lorises appear to be among the most specialized primates—and not only in the caudal region. All bony protrusions of the vertebrae in Lorisidae show a greater smoothness than do those of the other primates. Most of the projections are rounded and are not pointed at their tips.

An exception to this rounding off is found on some of the dorsal spines of the African lorisine species, *Perodicticus potto*. Starting at the third cervical vertebra and increasing in height on the following vertebrae, the dorsal spines become most prominent and pointed. The two distal cervical and the two proximal thoracic vertebrae are involved. In fact, here those spines are much higher than in any other of the nonhominoid primates. For instance, in pottos, the

Figure 6–4 *Tarsius*, third right rib, seen from above. Note the long articular surface (black).

highest dorsal spine in the second thoracic vertebra is twice as high as the vertebral body plus neural arch together. These dorsal spines protrude through the skin in adult pottos and are covered with a cornified epithelium. This peculiar structure was formerly believed to have a defensive function, and pottos were said to attack possible enemies with bent neck and to butt with the points of these spines (Figure 6–5). Later, Walker (1970) discovered that the neck bending behavior is not correlated with aggression at all but plays a role in nonagonistic social interactions. Walker observed pottos rubbing their necks in greeting ceremonies and other social contacts. Led by his observations, Walker looked at the histology of the covering epithelium in the potto's neck. He found that this epithelium is highly sensitive and equipped with tactile nerve endings. Now it appears obvious that these dorsal spines cannot be defensive organs. Looking at them with the Walker's explanation in mind, one can see that the spines are indeed totally surrounded and submerged into the dense fur of the animals and thus could certainly not be of much harm to a possible enemy. The covering epithelium also appears to blunt the points. The case of the spines of pottos beautifully demonstrates how important it is to know about an animal's behavior before attempting to explain its morphology.

Pottos also have many presacral vertebrae, and, compared with other prosimians, a comparatively barrel-shaped trunk. All four genera of the Lorisinae have dorsoventral foramina located in the

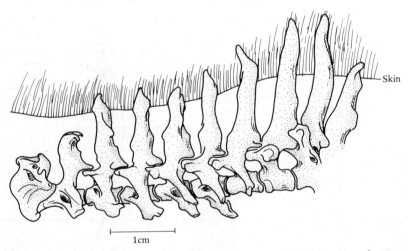

Figure 6–5 Neck vertebrae (1–7) and thoracic vertebrae (1–3) and spine of fourth thoracic vertebra of *Perodicticus potto* showing the extreme elongation of the dorsal spines.

bases of the transverse processes of the thoracic vertebrae (Figure 6–6). This feature distinguishes the subfamily from all other primates. The other African genus of this subfamily, *Arctocebus*, has peculiar ribs. These ribs are broad and shinglelike in their dorsal (vertebral) third, and they overlap each other just as shingles on a roof do. Jenkins (1970) compared these ribs with those of the sloths, which show the same type of rib morphology. However, here again, a lack of behavioral knowledge made it impossible to give a valid interpretation to these ribs, although Schultz (1961) believed that the overlapping ribs might give protection to the animal. Positioned inside the body-outline and only overlapping a small area of the back, the ribs could hardly prevent a fatal bite from behind as can a hard, interlocking surface (for instance, the hard surface in turtles and armadillos). Jenkins also could not find a satisfactory functional explanation for the peculiar ribs of *Arctocebus*. The two Asian Lorisine genera—*Nycticebus* and *Loris*—do not exhibit any such unique traits in the vertebral column. They are very much like

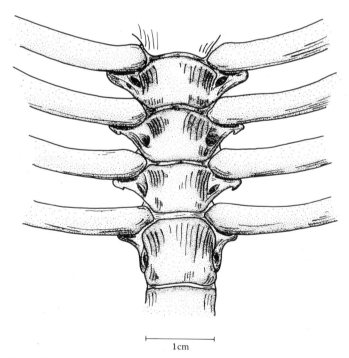

Figure 6–6 Ventral view of the thoracic vertebrae (13–15 and lumbar vertebra 1) of *Perodicticus potto* showing foramina typical for lorisiform prosimians.

African *Arctocebus* in not having pointed cervical processes, but they also have the foramina of the transverse processes in the thoracic region. All four Lorisine genera have reduced tails. Within the Prosimii this is also the case for the Malagasy genus *Indri*. All other prosimians have long tails.

The vertebral column of Anthropoidea does not exhibit many distinctive or unusual morphological features. Thus, the appearance of the vertebral column is very uniform in all the Cercopithecoidea. Partly at least, they show one obvious feature distinguishing their lumbar vertebrae from those of South American Cebidae and the Hominoidea. In cercopithecoid monkeys and most Ceboidea, the transverse processes insert directly on the sides of the vertebral body, usually at the place where its diameter is widest. In the larger cebids (*Alouatta, Lagothix,* and *Ateles*) and Hominoidea, however, these processes arise from the bases of the neural arch. There is variety in the vertebral columns of South American monkeys, whereas there is uniformity in the Old World monkeys (Figure 6–7).

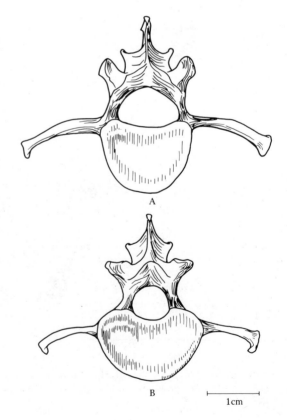

Figure 6–7 (A) Fourth lumbar vertebra of *Ateles* typical for New World monkeys and (B) fourth lumbar vertebra of a *Macaca* typical for Old World monkeys.

Erikson (1963) suggested that vertebrae of the thoracic and lumbar regions should be counted "functionally" and advocated not following the traditional subdivision into rib-bearing and non-rib-bearing vertebrae. Erikson thinks that the change of position of the articular processes is much more informative: the back and forth "spring" movement between vertebrae with lumbar positioned articulations involves up to three of those vertebrae that had usually been attributed to the thoracic region. Consequently, the position of articulations, not presence or absence of ribs, are tallied in Erikson's method of counting.

In all vertebrates the lumbar region is a rather rigid structure. The position of the articular processes prevents adjoining vertebrae from turning against one another. This is additionally reinforced by the accessory bone projections (*processus accessorii*) locking these articulations from below and from the sides. In some cebids the dorsal spines of this region are so long that little interspace remains between the individual vertebrae. Additionally, these spines are elongated into a point on their front end and split into a V-shaped point at the hind end. Thus, only slight dorsal bending of this region is possible because the cranial points of one vertebra become locked into the V-shaped caudal spine chamber of the vertebra in front of it. This condition could be related to the use of the tail as a fifth limb.

The sacrum of South American prehensile-tailed monkeys reflects the functional abilities of the adjoining tail region in different ways. In prehensile-tailed monkeys the contact surface of the sacrum with the iliac blades is known to be somewhat rougher than in other long-tailed primates, thus enlarging the area of interlocking contact between sacrum and pelvis; see Leutenegger (1974). The morphological differences between prehensile-tailed monkeys and long-tailed monkeys that do not use their tails as a prehensile limb are not as large as the differences in function. This can be easily understood: when hanging at the tip of the tail only, an action often exercised by these animals, the body weight is not transmitted to the pelvis but passes mainly through the axial skeleton.

The sacrum is also an indicator of tail length in primates, correlating with the length of the tail region. Thus, in long-tailed forms the sacrum is usually composed of three sacral elements. In many cases when tail length decreases, there is an increase in the number of sacral elements. Even so, this is not always the case in prosimians and monkeys with reduced tails. Hominoidea, which do not have an outer tail, all have sacra that are combinations of more than three, the generalized number of sacral vertebrae in primates.

In addition, one can observe that the neural arch tapers posteriorly to a very small diameter in the sacra of apes and man. Looking at the sacral regions of many living primates, one can see that there is quite a degree of difference between primate genera in the size of the neural channel of the sacra. Comparing the opening in front with the caudal opening of the sacral neural arch, we can observe the following: generally the diameter of the neural arch is triangular in the sacrum. In long-tailed lemurs, Callitrichidae, and Ceropithecinae, the cross-sectional area of the opening at the front is broader at the base than it is high, and the opening at the hind end remains nealy as high, but the breadth has decreased considerably. In Hominoidea the opening at the front end of the canal is like an equal-sided triangle; the caudal opening, however, is usually very small and much broader than high, if there is an opening at all, both the breadth and the height are reduced but the height is reduced the most. Both height and breadth are reduced in macaques and baboons together with some reduction of the tail region. In *Cacajao*—the only South American monkey that exhibits reduction of this part of the axial skeleton—the opening at the end of the sacral neural canal is low and broad, with the breadth reduced to about one third of the front opening (Ankel, 1972).

In monkeys with a prehensile tail, the opening at the end of the sacral canal is as large as or sometimes even higher (not broader) than the one at the front. This condition is also reflected in the height of the neural arches in the first tail vertebrae. The neural arches are much higher in South American prehensile-tailed monkeys than in any of the monkeys and prosimians with long tails that do not have a grasping function. In primates with no tail or nearly no tail there is also no neural arch above the few vestigial vertebral bodies remaining. Thus, about seven to eight of the first tail vertebrae are equipped with neural arches in the prehensile-tailed woolly, howler, and spider monkeys, whereas "common" long-tailed primates have only three or four tail vertebrae with neural arches. In the Hominoidea the neural canal ends within the sacrum. The size of the neural arch is correlated with the diameter and proportion of the spinal cord in that region. The cord is known to be very large within the sacrum in prehensile-tailed monkeys, and the canal holds a double system of arteries as well. Both the nervous system and the blood vessel system have to supply the highly sensitive area at the ventral end of the tail that is equipped with a well-developed tactile friction skin in woolly, howler, and spider monkeys. These primates also have very deep ventral arches in their tails and show another morphological expression of the greater functional capacity of the caudal region: the longest tail vertebra is positioned farther off the tail root than in other long-tailed monkeys, and the tail vertebrae at the tip of the tail are short and flattened.

Cercopithecoidea are amazingly uniform in the morphology of their vertebral columns. The tail is reduced to some degree in baboons and in some macaques.

The Hominoidea have several characteristics in common in the vertebral region that distinguish them from all other primates. In the cervical region, Pongidae have enormous dorsal processes, especially well developed in male orangutans and gorillas. These processes are not as long (relatively) as those of *Perodicticus potto*. They are only present in the cervical region where the dorsal spine of the seventh vertebra is usually the most prominent one. These spines are not pointed but end in a node. Enlargement of the dorsal spines in the cervical region is obviously correlated with the enormous bulk of neck musculature insertion at the back of the heavy head in these, the largest of primates. In man also, the seventh cervical vertebra has the longest dorsal spine. It can be felt below the skin and this vertebra was therefore named the *vertebra prominens*.

Farther back in the vertebral column of Hominoidea, the increase in diameter in the vertebral bodies is more apparent than the increase in length. The opposite is the case for all monkeys and prosimian vertebrae. Thus, especially in the lumbar region of Hominoidea, the vertebrae are short and very broad. They do not have the ventral keel that is found in the lumbar vertebral bodies of all the other primates except *Tarsius*. The sacrum is long and tapers at its caudal end, whereas the tail has been reduced to only a few tiny bony elements and is not visible externally in living Hominoidea. The articular surfaces with the pelvis and the sacrum in man are larger than in pongids of the same approximate body weight. It is thought that this enlargement has developed as a result of upright posture and locomotion, for we know that this is the area where the entire weight of the presacral body of man is transmitted through the vertebral column to the pelvis and thence to the legs.

The ribs of all Hominoidea are more heavily bent near their articulation with the thoracic vertebrae than in other primates (Figure 6–8). The transverse processes are at the same time angulated more dorsally. This angulation of the struts of the tuberculum of the ribs and the bending of the ribs result in a barrel-shaped, rather than narrow, rib cage in Hominoidea. The thoracic vertebrae thus come in a position within the rib cage, whereas in all

Figure 6–8 (A) First and seventh ribs of *Theropithecus gelada*, characteristic for monkeys, and (B) first and seventh ribs of *Symphalangus syndactylus*, characteristic of Hominoidea (articular surfaces black, both animals of about the same body size).

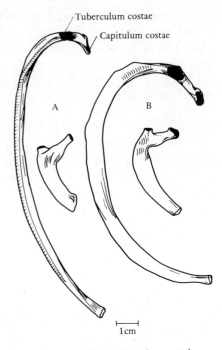

nonhominoid primates the less bent ribs are "hanging down" from the vertebrae. As already stated, a broadened breastbone always accompanies a broadened rib cage (Figure 6–9).

As we have seen, the breast-bone in all Hominoidea tends to be broad, and the breast bone of all monkeys and lower primates is very narrow (Figure 6–10). The ossification of the breast bones appears in seriated ossification centers within the cartilage and results in

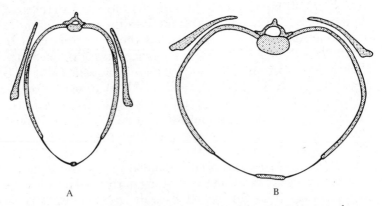

Figure 6–9 (A) Narrow "hanging" thorax, typical for prosimians and most monkeys, and (B) broad thorax, typical for all Hominoidea.

Postcranial Skeleton **167**

separate seriated bony elements, the so-called sternebrae. These bones remain separate throughout life in all prosimians and monkeys. Thus, their breast bones remain segmented, whereas, in contrast, these elements become fused among Honinoidea. Sternal fusion begins at the so-called sword process (xiphoid process or *processus ensiformis*). Among Hominoidea, late in development, this process ossifies but remains cartilaginous in other primates. The most proximal segment of the breastbone is broader than are the other sternebrae, and it is often broadest at the front end. The cartilage of the first pair of ribs inserts at the sides of this element. This element is called the *manubrium* or "handle" of the sternum because of its peculiar handlelike shape. Craniolaterally, this bone has two articulations with the inner ends of the clavicles. These articulations make up the only bony connection of the shoulder girdle with the axial skeleton.

The ribs are subdivided according to the place and manner of their connection with the breastbone. Most of the ribs, beginning with the first pair, insert with their cartilaginous ventral end between two adjoining sternal segments. These are called true ribs

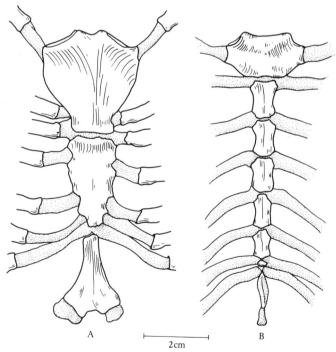

Figure 6–10 (A) Broad sternum, typical for Hominoidea, and (B) narrow sternum, typical for all other primates.

or (in the Latin expression, *costae verae* or also *costae sternales*). Some of the more caudally situated ribs join with their cartilaginous ends to insert at the distal end of the last sternal segments, so that the xiphoid process is between them. Together, the cartilaginous parts of these ribs make up the so-called costal arch. These ribs are therefore called arch ribs (or *costae arcuariae*). Two to three pairs of the most caudal ribs do not have any connection with the breastbone at all but are embedded in the musculature. These are so-called free ribs (or *costae fluctuantes*).

Ribs in basic structure are long narrow and flat bones; they are bent at the vertebral end in two directions: There is a torque within and also slightly along their long axis. True ribs have two articular facets at the vertebral bodies and another contacting the transverse process caudal to these. The heads of the true ribs usually articulate at the dorsal end of two adjoining thoracic vertebrae; thus, they are partly situated over the intervertebral disc. Farther back in the thoracic region, with the transverse processes of the ribs becoming shorter and the vertebral bodies increasing in length, the *tuberculum costae* and its contact with the transverse process is successively reduced. The more distal ribs are only connected to the thoracic vertebral bodies with their heads. In the same process, the articulation of the heads of the ribs moves on to the side of the more distally situated vertebral body. The first ribs are much shorter than the following pairs, and they are not twisted longitudinally. From front to back, the ribs quickly increase in length, remain more or less of the same length in the middle of the thorax and increase the length again distally. The free ribs are usually short; typically, they are not twisted around their longitudinal axis and are but slightly bent. The hindmost ribs may actually fuse into one or both sides of the vertebral body, thus producing an elongated transverse process-like structure located in just the same position as these processes have in the lumbar region. In concluding this account of the vertebrae, it is well to say that many studies have shown that the numbers of vertebrae vary intraspecifically in all primates. This variability least frequently affects the number of cervical vertebrae and is most often expressed in the caudal region.

SHOULDER GIRDLE

We have seen that the vertebral column and thorax of many non-hominoid primates are unspecialized and thus not very different from that of other quadrupedal mammals—for example, the cat. We also have reviewed how the extant members of the Hominoidea differ significantly from other primates in the morphology of the axial skeleton and in the shape of their rib cage. Among these animals and man, as has been stated, the thorax is broad and shallow, and the vertebral column is positioned within the outline of the rib cage, rather than at the utmost dorsal edge of the thorax. The real reason for the appearance of barrel shaped trunks in hominoids is not well understood. Nor do we know the time when this broadening first arose in the evolutionary history of the hominoid primates. There are different factors that could have influenced the original development of broad rib cages. One such factor could be large absolute size and concomitant increase in relative mass of the respiratory system and viscera. This factor could not apply to those few small prosimians, such as *Tarsius*, which also show chest broading. Another cause for the change in chest shape could be sought in diet. We know that mammals that live exclusively on leaves or grass tend to have broader trunks than those that subsist on high protein diets (e.g. carnivores). There are, however, primates that are specialized for a leaf diet—for example, the South and Central American howler monkeys, which have narrow and deep trunks. A third factor could be locomotion. Primates with barrel-shaped trunks not only have broad and shallow rib cages, but also this broadness affects the pelvis (broad ilia). This condition can in turn be influential for the position and function of the hindlimbs. However, we want to evaluate the influence of the broad thorax on locomotion and forelimb movement, and perhaps it relates to improved climbing. Only primates with barrel-shaped trunks have a more or less flat back and chest. In those primates the shoulder blades are positioned on this back and not at the side of the rib cage as in mammals with narrow trunks. In the latter the shoulder blades are positioned laterally. Thus, the shape of the trunk critically influences the morphology and function of the entire forelimb girdle.

All primates retain clavicles that are lost in many other mammals—for example, cats, dogs, and horses. As we have seen, clavicles are the only remnant of the exoskeletal elements derived from early vertebrates, and the clavicle is the only dermal bone in the primate postcranial skeleton. The clavicle attaches medioventrally at the manubrium stem and laterally with the acromion of the shoulder blade (scapula). Thus, the clavicle acts as a strut holding the shoulder blades apart. We have also noted that the clavicular articulation with the breastbone is the only attachment of the upper limb girdle to the post cranial skeleton. In those mammals that have lost clavicles, the upper limb girdle is embedded and attached to the trunk solely by musculature. In animals with clavicles, the upper limb girdle is attached mainly by the muscles that both move it and hold it in place at the same time. This condition gives in turn a high degree of mobility to that region. It is the function of the clavicles to keep the scapulae, and thus the upper arm articulation, in a lateral position or, in the case of animals with a broad trunk, in a dorsal position. This condition gives a higher degree of mobility to the articulation of scapula and humerus, and thus the entire upper extremity—a feature that is of great importance to animals that live in a three dimensional environment (Jenkins et al., 1978). Primates in particular, with their broad trunks and dorsal scapulae, have a three-dimensional area of forearm function as compared to, for instance, the horse, in which the forearm movement is restricted to an anteroposterior parasagittal plane that is essentially two dimensional. In sum, the presence of clavicles in the primate shoulder girdle is of critical importance for many forelimb activities that are, in part, responsible for the high adaptability of the primates. In barrel-trunked primates, the clavicles are longer than in primates with narrow trunks and are also directed cranio-laterally rather than laterally as in lower primates.

The shoulder blade is a more or less triangular, flat bone. The blade has a cranial, medial, and a lateral edge. The lateral upper angle enlarges or spreads out around the concave glenoid fossa—the articular facet for the globular head of the humerus. From this angle to the vertebral edge of the scapular triangle runs the *spina scapulae*, or scapular spine, an elevated bony ridge that flattens out toward the vertebral edge of the bone. The spine becomes robust and broad near the glenoid fossa where it is produced into a broad and strong bony process, the so-called acromion. The spine is somewhat tilted in a ventral direction, forward and medially; it thus overlaps the humeral head where it meets the lateral end of the clavicle. The *spina scapulae* divides the shoulder blade into two planes, called *fossae supra-* and *infraspinata*. The two fossae are extremely variable in shape, both intra- and intergenerically. In addition, the outline of the entire bone is variable within and between different primate genera (Figure 6–11).

The shoulder blade has a framelike construction. The blade is thick at its edges and in the area of the spine. The spine of the shoulder blade not only serves as a bony protrusion for the attachment of muscles but also stiffens the entire bone (this is according to the T-beam principle). The planes of the shoulder blade are very thin in their centers, especially the large subspinal plane of the largest primate shoulder blades—for example, those of *Gorilla*

Figure 6–11 Right shoulder blades of (A) *Lagothrix*, (B) *Theropithecus*, (C) *Symphalangus*, and (D) *Pan*.

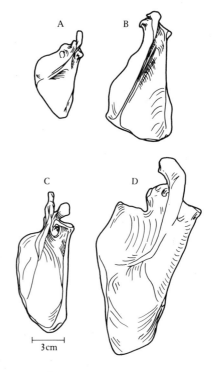

or man. The parts that are subject to forces are the edges: the scapular spine and the glenoid, acromial, and coracoid areas. The coracoid is a bony process that originates medially to the glenoid on its cranial edge. This bony process tilts in lateral direction in front of the head of the humerus and the glenoid fossa. The acromion, coracoid, and clavicle are surrounded by strong ligaments that form part of the articular capsule. Numerous muscles attach at the edges, the spina, the acromion, and the coracoid process of the scapula. These muscles move it upon the rib cage and hold it in place. Thus, the shoulder blade "hangs" in an envelope of muscle. The upper limb shifts its position in unison with the scapula and consequently, the length and proportions of the scapulae act as lever arms and are important for the degree of mobility of the upper limb. The shape and proportion of primate shoulder blades are correlated with different types of forelimb use. Shoulder blades of gibbons, for example, are elongated and narrow. The scapulo-humeral articulation is free and mobile, the glenoid fossa shallow and small. The entire articulation is secured mainly by tendons and musculature. This articulation of gibbons is thus highly mobile but not very stable.

In all the barrel-chested primates the position of the shoulder blade is different from that of the primates with narrow chests. The high position of the shoulder blades dorsally on the broad trunks of hominoid primates results in an exposed articulation of the upper arm. The possible range of movement of the upper extremity is thus higher than in non-hominoid primates. The fore-extremity can be envisioned as moving along the radius of a circle, the center of which is the scapulo-humeral articulation, it radius the entire length of the forelimb, plus hand length. Because the scapula is not fixed or anchored down, the center of this space globe is also mobile. The efficiency of this structure depends on the absolute size of the animal involved and on its forearm length. A variety of factors may influence size and forearm length. Increased body size is accompanied by an enlargement of all inner organs. It also requires an increase in bulk of musculature and these, in turn, need more insertion surface. More forelimb use can also result in enlargement of the muscle bulk. Even diet possibly influences the dimensions of the alimentary tract and thus may influence the shape of the trunk. Locomotion, posture, and pelvic breadth can also be of importance to trunk dimensions.

FOREARMS

The broadening of the trunk is also accompanied by a change of the relative position of the upper and lower articulations of the humerus. This condition is called humeral torsion. Such medially directed torsion is nearly 80 degrees in man, about 30 degrees in *Hylobates*, and only 10 degrees in *Macaca*. Without such torsion within the humeral shaft, the hands of hominoids would always be turned outward when in a relaxed position. To say it differently, if the lower articulation of the humerus is positioned perpendicular to the long axis of the body, the head of the humerus faces backward in a macaque but inward and backward in homonoid primates (Figure 6–12). In some primates a foramen is found at the inner side of the outer end of the humerus. This foramen is a survival from archaic mammals, and has the complicated name "entepicondylar foramen." Where it is present, the median nerve and the brachial artery pass through it. This foramen has been retained in most living prosimians except the two African lorises *Perodicticus potto* and *Arctocebus calabarensis*, where it is only variably present. Species of many South American monkey genera also have this foramen— for instance, in *Saimiri, Leontocebus, Pithecia* (Figure 6–13), *Cacajao*, and sometimes *Cebus* and *Aotus*. This foramen is not

Postcranial Skeleton

Figure 6–12 (A) Head of right humerus of a *Macaca* and (B) head of right humerus of man. Line indicates the position of the axis through the elbow joints; the arrows indicate the center of articulation with the shoulder blade (A and B reduced to the same size). (C) Caudal view of the right humerus of a *Macaca*, position of the humeral head typical for a quadruped, facing backward and (D) dorsal view of the right humerus of man, articular head facing dorsomedially (C and D reduced to the same length).

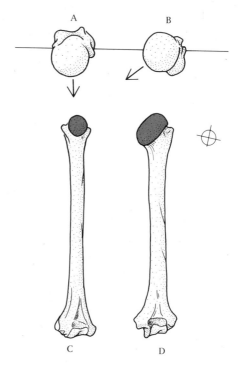

Figure 6–13 Dorsal view of the right humerus of *Pithecia* with entepicondylar foramen (arrow indicating foramen).

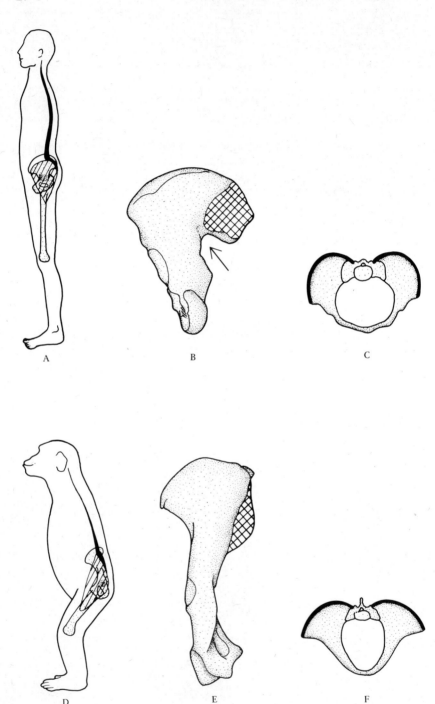

normally found in *Cebuella, Callithrix, Ateles, Alouatta,* and *Callicebus* nor among any living catarrhine primates. It is said that the entepicondylar foramen occurs in rare or exceptional cases among most primates, sometimes even in *Homo.*

The forearm contains two bones (radius and ulna) from which both mobility and stability of the forearm originate: they are positioned parallel to each other when the volar (inner) surface of the hand is directed upward, and the thumb points laterally. This position of forearm and hand is called supination. In this position the radius is the lateral element, and the ulna is positioned medially. If the volar surface of the hand looks downward, the thumb lies medially (pronation), and the radius then crosses over the ulna: the upper head of the radius rolls slightly inward on its articular facet with the ulna when the hand is pronated, and it rolls in the opposite direction when the hand is supinated (Figure 6–14). This high rotary range of the forearm is of importance for primates

Figure 6–14 Diagram comparing (A) man with (D) *Gorilla* and (G) *Macaca,* showing the vertebral column (black), hip, femur and (hatched) the approximate position of the gluteus maximus muscle. Right hipbones of (B) man, arrow indicating incisura ischiadica, typical for man; right hipbone of (E) *Gorilla* both seen from inside, cross-hatched the area of articulation with the sacrum; left hipbone of (H) *Macaca* seen from the back to show ischial callosity (gray). Pelvis of (C) man seen from above; pelvis of (F) *Gorilla* and pelvis of (I) *Macaca* seen from the front to demonstrate the striking differences in the way the hipbones are curved (outlined in black). (Sizes are not proportional.)

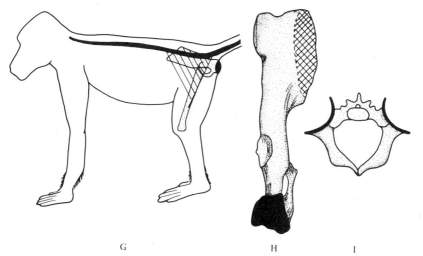

that live in trees. A primate hand has to be able to grasp branches and to hold on in many different positions. Thus, the rotary mobility of the forearm adds to the faculties of the upper extremity. In man and apes the distal end of the ulna is reduced to a narrow styloid process and does not articulate intricately with the carpals (triquetrum and pisiform) as it does in monkeys. This feature was believed to be typical for hominoid primates. Cartmill and Milton (1977), however, have shown that these features of the wrist are also found in lorisiform prosimians, and thus, they cannot be interpreted as being correlated with armswinging locomotion, as had been done. We have seen the fundamental differences in form and function of the two limb girdles. The shoulder girdle is almost totally devoid of any bony connection with the trunk and is highly mobile. Contrary to this, the hindlimb girdle appears rather rigid and designed for stability, transfer of propulsive forces, and weight bearing purposes. A similar phenomenon can be observed in the stability of the hindlimb generally.

Adaptation to terrestrial living usually reduces range and flexibility of the limb movements. Thus, for example, in horses who move with mainly anteroposterior excursions of the limbs, the clavicles are reduced, and ulna/radius and tibia/fibula respectively are fused together into single elements, as are many of the foot bones. In broad-chested primates the breastbone is also broad and the clavicle relatively longer than in narrow-chested primates. The clavicle of hominoid primates also describes an S-curve, in order to follow the shape of the rib cage.

PELVIC GIRDLE

The hindlimb girdle differs fundamentally from the shoulder girdle, for it has an extensive bony connection to the axial skeleton. The sacral region, which is the longest element within the vertebral column (being a combination of three or more vertebral segments), acts as keystone between the two iliac blades. The transverse parts of the sacrum are enlarged laterally and have a rough dorsolateral surface, fitting onto the corresponding ventral surface of the iliac blades. Both these connecting surfaces are bound together with strong, tightly applied ligaments forming an immobile capsule, the so-called "amphiarthrosis." This type of "articulation" is not constructed for mobility but rather for stability and absorbtion of shock. Among some primates this amphiarthrosis can even ossify in old age. The pelvis is formed by a pair of hipbones. They meet ventrocaudally at the sacral articulation in a symphysis or a cartilaginous junction that is reinforced by ligaments. This normally tight symphysis can temporarily be loosened somewhat; for example, during the process of birth in humans, when the large headed young to be born has to pass through the bony birth canal made up by the sacrum and hipbones. The symphysis only ossifies late in life in primates. The symphysis is said to be permanently open (that is, unfused) in the prosimian species *Loris tartigradus.*

Each hipbone (*os coxae*) is a combination of three elements that ossify independently and fuse when primates reach adulthood, after which time the three parts of the bone can no longer be recognized. The three elements forming the hipbone are the ilium, ischium, and pubis. The iliac blades form the proximal part of the pelvis that at its base includes about a third or so of the articular fossa or acetabulum, which holds the femoral head. The ventral bony part—including the symphysis—is called the pubis, and dorsal to the pubis lies the ischium. Both pubis and ischium form the lower part of the acetabulum. Pubis and ischium enclose a large paired opening behind the acetabulum on each side, the so-called obturator foramina. Those openings are covered by ligamentous tissue in living animals. The point where the ilium, pubis, and ischium meet is roughly positioned in the center of the acetabulum. As is well-known, the pelvis surrounds the birth canal of females. The pelvis is also subject to forces that originate from the locomotor activities of the hind legs and partly derives stresses from resting postures (sitting). These several functions act together to create different demands on the pelvic morphology. This condition is especially true in those primates among whom the inner diameter of the female pelvis and the size of the full term newborn offspring are critically close to each other—for example, in *Homo sapiens.* In such primates the infant may be unable to pass through during labor and both mother and offspring may die. Only we can remedy this problem with surgery (cesarean section). This crucial "bottleneck" situation also exists, however, in certain nonhuman primates that combine single births, highly developed brains, and newborns relatively large in correlation to the body size of the mother (Leutenegger, 1970). Schultz (1962) earlier reported such birth difficulties in colonies of *Papio hamadryas.* A pelvis with an outlet too narrow for the full term fetus would be highly selected against, but the alternative, selection for smaller brain volume would be even more nonadaptive.

In many prosimians the ilium is narrow, round in cross-section, and rod-shaped. In monkeys it is commonly an elongated blade that is bent along its long axis in such a way that it is concave dorsolaterally. The pelvis is broad and very long in pongids and to a lesser degree in hylobatids, the ilia forming flat blades in all apes. The ilium is short, broad, and bent ventrally in hominids. In man the articulation between pelvis and sacrum is also larger, both relatively and absolutely than in any other primates. These changes narrow the space between the sacro-iliac articulation and the hip joint, and form a distinct indentation (as seen from the side) in the sacro-iliac articulation area (see Figure 6–14). This angle is typical for hominids and is called the *incisura ischiadica*, or greater sciatic notch. All these changes in the human pelvis are correlated with the comparatively large size of the extensor musculature in the hip articulation. Especially, the *musculus gluteus maximus* is shifted in relative position to this articulation in upright man; it inserts in front of the center of the hip articulation in quadrupedal primates and extends over this articulation in man. The *musculus gluteus maximus* also changes its function between the two, being mainly a flexor and abductor in the hip joint of lower primates and a very powerful extensor in bipedal man. In all Old World monkeys the ischium is greatly enlarged dorsocaudally. These enlargements are covered by ligamentous and corneous tissue and protrude through the fur. Catarrhine monkeys sit on these pads or "ischial callosities" that are often large. The shape of the ischial callosities varies significantly between different species. Humans, not having ischial callosities, sit on a cushion composed of the gluteus maximus plus subcutaneous fat. Small ischial callosities are found in rare cases in both the lesser and the great apes. The ventral, cranial, and caudal edge of the articular fossa for the femoral head is strengthened by a semicircular thickening of the bone. Variations in the ventral length of the pubis and the outline of the obturator foramen show sexual dimorphism in species with narrow birth canals and large-headed newborns. As already stated, examples of this include humans and the South American squirrel monkey *Saimiri*.

  The proximal segment of the hindlimb contains one rigid bone, the femur. At the inside of the upper end, the femoral head projects from an upward angled femoral neck. Neck length and angle vary with differences of pelvic morphology and locomotion. The femoral head is of globular shape in most primates except in the highly specialized leaping prosimians, Galagidae, and Tarsiidae. In these animals the femoral head tends to be cylindrical (with the long axis of the cylinder perpendicular to the long axis of the femur). In many of the more terrestrial primates, the femur is bent craniocaudally. In quadrupeds, the two condyles of the distal articulation are equal in size. Whereas in man, with his comparatively broad pelvis and considerable distance between the pelvic articulations of the two sides, the thighs tend to converge downward, but the opposite is true of pongids. Consequently, the medial or inner trochlear condyle of the lower femoral articulation is larger in man, and the lateral one is usually bigger in pongids. The femoral shaft is narrow, slender, and straight in lesser apes, and it is very robust and bent forward in pongids. In hominoids also, the lower end of the femur has a deep groove for the patellar tendon.

In bush babies and tarsiers, which move by leaping, the femoral head and hip articulation are cylindrical rather than globular, as we have stated, and here also, the long axis of the femoral head stands more or less perpendicular to the long axis of the femur. In man the femoral head is relatively large, even compared to the proximal articulation of the humerus. The opposite size proportion between these articular heads is found in gibbons, which move predominantly with the arms. In quadrupedal primates the diameters of humeral and femoral heads are subequal.

HINDLIMBS

Femur, tibia, and fibula are the three long bones of the hind leg. The femur, most proximal and largest of the three, shows three prominent bony protrusions at its upper end. These are: 1. the inward-directed and more or less ball-shaped *caput femoris* or femoral head that articulates with the hipbone. 2. at the lateral side of the femur's upper end is located the second protrusion—the *trochanter major*— or greater trochanter; on this insert the gluteus medius and minimus muscles that, in man, abduct the thigh and rotate it inward as well as some lesser functions. 3. Below the femoral head and medially to it, we find a bony protrusion on the hinder aspect of the femur that is called the lesser trochanter or *trochanter minor*. Sometimes, a third trochanter is present (*trochanter tertius*): this is a small elevation laterally on the shaft below the *trochanter major*. The trochanter may also be called *tuberositas glutea*. On the front both the greater and lesser trochanter are connected by a bony crest that is called the intertrochanteric line.

On the backside of the femoral shaft is another bony crest, the *linea aspera*. This is the line of insertion of several important muscles. At the knee the femur ends in two more or less rounded knobs, the outer and inner condyles. The axis through the center of these articular knobs lies approximately at a right angle to the long axis of the femur. This angle varies somewhat in characteristic fashion in different primates, as we shall soon see. The condylar articulations enclose between them the facies patellaris on the front face of the bone and contact the upper articular facet of the adjoining long bone below, the tibia.

On the upper facet of the tibia, the two articulations with the femur are separated by a bony prominence, the so-called *eminentia intercondylaris*. Medially on the front of the tibia, a more or less roughly sculptured ridge—the tibial tuberosity—brings about a triangular cross-section of this bone that farther back fades out into a more circular shaped cross-section. At the lower end the tibia extends on the inside into a pronglike bony extension: The medial malleolus that extends over the talus of the foot. Looking at the underside of the upper lateral condyle protrusion, one finds a small articular facet that accommodates the head of the third lower long bone, the fibula, which is positioned outside the tibia. On the backside of the tibia, an oblique ridge crosses the shaft, beginning outside at the upper end under the lateral condyle, extending over about one quarter of the shaft length, and ending on the inner face of the shaft. This oblique ridge is accurately termed the *linea obliqua*. The fibula in its entirety is thin and slender, for it is not a significant weight bearing element of the leg. This bone is connected to the tibia by means of two only slightly mobile articulations (synovial articulations) on both its upper and lower ends. Both ends of the fibula are enlarged. The upper end of the fibula is called the fibular head (*caput fibulae*). This head is usually positioned slightly behind the tibia and does not have any connection with the knee joint. Its attachment to the tibia is secured by ligaments in front and in back. On the inner aspect of the fibular head, we find an articular facet that contacts with the tibia. The lower end of the fibula is larger than its head and protrudes down as the lateral malleolus on the outside of the ankle. On its inner aspect a large, rounded articular facet contacts the corresponding facet on the talus of the foot. The shaft of the fibula has a triangular cross-section in most primates.

In 1972 an extensive evaluation of the long bones of the anthropoid hindlimb was published (Halaczek, 1972). We can review some of the results of this study, as follows.

It appears that among New World monkeys the major factor that influences the morphology of the long bones of the hind leg is absolute body size and not the mode of locomotion. Two different groups of hind legs can be distinguished. They are called, by

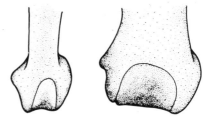

Figure 6–15 Facies patellaris of (left) *Saimiri* and (right) *Alouatta*, demonstrating the differences between platyrrhines of small and large bodysize in this respect.

Halaczek—the *Aotus*-group—containing Callitrichidae and cebids of small body size as well as *Pithecia* (the genera *Chiropotes* and *Cacajao* were not available for this study) and the *Ateles*-group—covering *Ateles, Brachyteles, Lagothrix,* and *Alouatta,* the South American monkeys of larger body size. Interestingly, genus *Cebus* does not fit well into either of the two groups but exhibits an intermediate morphology. In the group of small-sized species, the skeleton of the hind leg is shorter than their skeletal trunk length, and in the large-sized group (excepting *Alouatta*) the skeleton of the hind leg is longer than the skeletal trunk. In the small-bodied group, the femur is typically shorter than the tibia, and in the large-sized group the femur is always longer than the tibia. Among small South American monkeys, there is a third trochanter that is always missing in the large-sized platyrrhines. The *Aotus*-group shows a tendency to have a convex medial bending of the femoral shaft. Contrary to this, the femoral shaft in the *Ateles*-group appears to be convex laterally when viewed from the front. The *facies patellaris* is always higher than broad in the *Aotus*-group (Figure 6–15), and it is almost (but not quite) as high as it is broad in the *Ateles*-group. The angle between tangents on the distal condyles of the femur is open laterally in the small monkeys and open medially in the large group. Moreover, there is an inward torsion in the femur of the small South

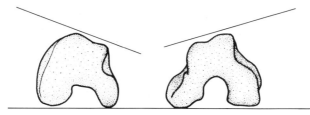

Figure 6–16 Lower ends of right femora of (left) *Callithrix* and (right) *Ateles*, showing the tangents enclosing the condyles, open laterally in the platyrrhine group of small body size and open medially in the platyrrhine group of large bodysize.

American monkeys and an outward torsion in the larger kinds of monkeys. The upper portion of the fibular diaphysis is positioned laterally and behind the tibia in the smaller-sized group, and in the large-body-size group it is positioned laterally to the tibia.

Halaczek also concluded that in basic construction the three long bones of the hind leg in New World monkeys cannot easily be distinguished morphologically from those of Old World monkeys.

Among Old World higher primates the morphology of the three hind leg bones exhibits a characteristic combination of features in each of the four families (Cercopithecidae, Hylobatidae, Pongidae, and Hominidae).

In Cercopithecidae the length of the hind leg bones—that is, femur plus tibia—is shorter than the length of the skeletal trunk with one exception, namely the proboscis monkey, *Nasalis*. Femur and tibia are about 25 percent longer than the skeletal trunk in Hylobatidae. Among Pongidae we find that in the orangutan, hind leg length and trunk length are about equal, whereas leg length slightly exceeds trunk length in chimpanzee and gorilla.

In *Homo sapiens* leg length always exceeds the length of the trunk. Cercopithecidae typically have a femur and tibia of about equal length, but the tibia is sometimes slightly shorter than the femur.

The hind leg bones of gibbons are characterized by the fact that the tibia is—as a rule—shorter than the femur. This condition is also the case in the great apes and humans.

Commonly in orangutans, chimpanzees, and ourselves, the head of the femur extends above the greater trochanter.

Concerning the robusticity of femur and tibia, we find that both these long bones are about equal in robusticity among Cercopithecidae, and in the gibbon family both femur and tibia are rather slender and the tibia is thicker than the femur. Contrary to this condition of the lesser apes, these leg bones are rather robust among Pongidae, and in the latter family the thickness of the femur exceeds that of the tibia. In *Homo* the robusticity of femur and tibia is less pronounced than in pongids, but the human tibia exceeds the femur in robusticity.

With the exception of some colobines, the trochanter major extends up higher than does the femoral head in ceropithecines. Contrary to this, the head of the femur is elevated above that of the trochanter major in Hylobatidae.

Both caput femoris and trochanter major are of equal height in the gorilla, whereas in the chimpanzee, the orangutan, and ourselves the caput femoris is higher than the trochanter.

Frequently, Ceropithecidae have the smooth articular surface covering the femoral head extended onto the hinder aspect of the neck, and the articular surface does not exceed beyond the caput femoris in both the lesser and the greater apes. In contrast, this same articular surface extends on the anterior aspect of the femoral neck in *Homo sapiens.*

The angle between the long axis of the femoral shaft and that through the femoral neck and head is about 120 degrees in cercopithecids, and it usually measures about 130 degrees in the lesser apes and the chimpanzee, whereas it averages 127 degrees in gorillas, 140 degrees in orangutans, and 131 degrees in man. New World monkeys show modification of these proportions in the Atelinae, where both fore- and hindlimbs are elongated and the skeletal trunk length relatively shortened.

Among both lesser and greater apes the forelimb is greatly elongated, most markedly in *Hylobates* and *Pongo.* In contrast, a pronounced elongation of the hindlimbs typifies bipedal man. Extreme elongation of limbs within Order Primates is only found in species with the most specialized locomotor behavior—that is, with the leaping, the armswinging, and the bipedal species.

In correlation with the increased weight-bearing demands on the hindlimb in man, all the articular surfaces of the hindlimb long bones, and of the talus and calcaneus of the foot, have increased in surface area. Likewise, size enlargement of the articular surfaces on arm bones can be seen in those primates exhibiting forelimb elongation and forelimb preference during locomotion. This includes both knuckle-walkers (the African great apes) and the true brachiators, namely the lesser apes. Although in cercopithecids and lesser apes the shaft of the femur shows no torsion, it does exhibit a considerable outward torsion among large apes. In humans a high degree of inward torsion is the rule.

In both cercopithecids and men two tangents drawn along the anterior and posterior aspects of the distal condyles of the femur (Figure 6–17) will usually diverge laterally. Both these condyles are equal in size in lesser apes and thus, the tangents do not enclose an angle but run parallel to each other. Contrary to this, the great apes exhibit a large medial condyle, and the angle between the two tangents already described opens medially as a rule. In all the Old World monkeys and apes, the lateral part of the lower articular surface of the tibia tilts in a proximal direction. This articular surface is flat in its entirety in humans and is always positioned at a right angle to the long axis of the bone. With Old World monkeys and apes we find that the tibia shows a distinctive inward (medial)

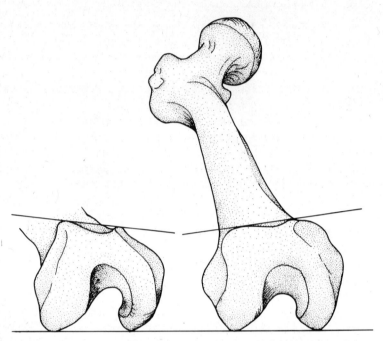

Figure 6–17 Lower end of right femora (reduced to the same size) of man (left) and *Gorilla* (right) showing the tangents enclosing the condyles.

torsion. This torsion is especially developed among Hylobatidae, *Gorilla*, and *Pongo*. In *Homo* there is a characteristic tibia torsion that is always directed laterally. A fibula that is shorter than the tibia is characteristic for Cercopithecoidea, Hylobatidae, and Pongidae. Only in Hominidae the fibula is sometimes found to be longer than the tibia.

With unspecialized primates the hind extremities are usually longer than the fore extremities and shorter than the skeletal trunk length. Femur and tibia are of equal length. These general proportions are found in both unspecialized New World monkeys and among Cercopithecoidea of small body size. They also occur in both the terrestrial *Erythrocebus* and the arboreal Callitrichidae.

HANDS AND FEET

Both hands and feet of primates pass through an early ontogenetic stage, that is reminiscent of an unspecialized tetrapod. From this early pattern (which has fifteen hyaline central elements) hands and

Postcranial Skeleton **185**

feet have evolved differently; see Figure 6–18 and Table 6–2. Respectively, adult primates retain either eight or nine carpal elements, whereas the tarsus of such adults always consists of seven different elements. Among three prosimian genera *Indri, Lepilemur,* and *Avahi* (and in the hominoid genera *Pan, Gorilla,* and *Homo*) we find, that the central bone fuses with the navicular early in individual growth or is reabsorbed. Because of this fusion, species of these six primate genera have only eight carpal elements while all the other primates have nine. In Table 6–2 the different sequences of ontogenetic development into hand and foot of primates are indicated.

Relative differences in lengths of fingers and toes are usually expressed in the so-called formulae. These formulae simply give the

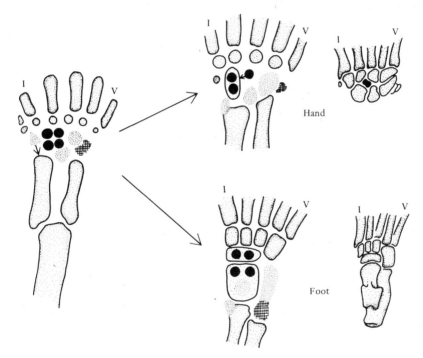

Figure 6–18 Phylogeny of primate hand and foot. (Left) unspecialized extremity, initial stage for both, hand and foot. (Above middle) diagram of primate carpals and (below middle) primate tarsals. (Above right) primate carpus and (below right) primate tarsus. Carpals and tarsals = white; praepollex = big dots; proximal elements (radial/tibial, ulnar/fibular) = fine dots; pisiform = medium sized dots; centrals = black. (See also Table 6–2.)

Table 6–2 Phylogenetic Derivation of Hand and Foot in Primates

Unspecialized Extremity	Primate Hand (Carpus)	Primate Foot (Tarsus)
Carpals/tarsals 1	trapezium	medial cuneiform
2	trapezoideum	intermediate cuneiform
3	capitate	lateral cuneiform
4	hamate	cuboid
5	reduced	reduced
2 proximal centrals	———	form talus together with intermedium
2 distal centrals	———	navicular
2 medial (radial) centrals	scaphoid	———
2 lateral (ulnar) centrals	proximal ulnar central reduced distal ulnar central separate or incorporated into scaphoid as in *Indri Lepilemur Avahi Pan Gorilla* and *Homo*	———
Intermedium	lunate	forms talus together with 2 proximal centrals
Ulnar/fibular Pisiform	ulnar forms triangular pisiform	fibular and pisiform form calcaneus
Radial/tibial	fuses with radius	fuses with tibia

Data from Steiner (1951).

number of the longest digit first, that of the shortest last, and all the others according to their length. Thus, the digital formula of the human hand is usually

3 - 4 - 2 - 5 - 1   or

3 - 2 - 4 - 5 - 1

that of the human foot

        1 - 2 - 3 - 4 - 5   or, less commonly

        2 - 1 - 3 - 4 - 5.

    A few primates have webbed hands or feet, that is, some of their fingers and toes are connected by variably developed skinfolds. Thus, in the prosimian genera *Propithecus* and *Indri* certain fingers and toes are webbed together. Species of *Prophithecus*—the sifaka—show webbing extending out toward the first articulation between second, third, and fourth toes, and here also, a small skinfold is found between fingers 3 and 4. In the *Indri indri* (babacoot or giant lemur), the webbing is more extensive than in *Propithecus* and certainly prevents fingers and toes 2 through 4 from moving independently: moreover, the membrane extends out further than the first joint between all toes and fingers 2 through 5. Among some species of *Cercopithecus* slight webbing of toes 2 through 5 and hardly recognizable webbing between fingers 2 through 3 can be found. In *Nasalis* toes 2 and 3 are usually connected by skin, and the lesser ape species *Symphalangus syndactylus* has taken both its generic and species name from the fact that its toes 2 and 3 are closely connected to each other by a skinfold. Such webbing is also frequently found in members of the related genus *Hylobates*. Gorillas often have toes 2 through 5 webbed out to the first joint as well as fingers 2 through 5.

    Primates, with few exceptions, typically retain unspecialized hands with five digits. The tips of fingers and toes are equipped with either claws or nails in primates, and moreover, nails are only found among primates. Even so, as noted in Chapter 2, not all primates have nails on all fingers and toes. Claws are hornlike structures that cover the outermost bone of either toes or fingers. It is characteristic of claws that they are compressed from side to side and curved lengthwise. In addition, the underlying outermost toe bone or terminal phalanx has the same shape and thus is an image of the overlying claw. Claws have two layers: a deep, typically thick layer (deep stratum), and a covering layer that is considerably thinner, hardened, and that functions as a protective envelope (superficial stratum). Nails are flattened from inside to outside and usually are made up of the superficial stratum only. Terminal phalanges of nail-bearing primates are broad and flat and are, in effect, a mold of the covering nail, just as a terminal phalanx carrying a claw is a mold of the claw. Claws extend beyond the phalangeal tips and end in sharp and pointed hooks, useful for many activities including climbing, digging, and scratching. In contrast, nails usually do not extend beyond the tips of the fingers and toes (Figure 6–19).

All members of Tupaiiformes have hands that are equipped with claws, and all five digits are positioned close together. Nevertheless the tree-shrew pollex can be spread slightly away from digits 2 to 5. Even though this digit in tupaias can be abducted to some degree, it cannot be put into opposition to the rest of the fingers.

The carpo-metacarpal joint (articulatio pollicis) of primates will be considered in detail because many authors think that the thumb and its independence from the rest of the digits in the hand of humans must have played a major role for the emergence of man as a manipulative, civilized creature with the ability to create culture; see Biegert, 1963. Hands and thumbs are of great importance for many activities of primates.

Even in non-primate tree shrews or the small-sized marmosets, grasping with the hands habitually involves all five clawed digits drawn together in unison. Among more advanced primates, the ability to spread the fingers and to separate the thumb from fingers 2 through 5 or even bring it into opposition to them becomes increasingly important and structurally more elaborate. However, the degree of manipulative ability is not solely reflected by the kind of articulation the thumb has with the hand. Thumb length, finger length, and tactile abilities provided by elaborate nervous supply are equally influential for the different kinds of manipulative abilities we find among primates.

It was Napier who defined the different manipulative properties of primate hands. According to Napier (1961, Napier and Napier

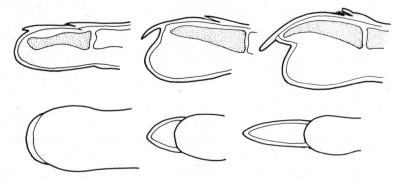

Figure 6–19 Fingernails and claws: (left) human nail; (middle) nail of a monkey; and (right) animal claws.

1967), primates show three kinds of grasping with their hands. These, mainly determined by the independence of the thumb movements, can be listed as follows:

Group 1. Non-opposable thumbs
Group 2. Pseudo-opposable thumbs
Group 3. Opposable thumbs

Major groups of primates may be ranked in these groups as follows:

Group 1. tarsiers, marmosets
Group 2. all prosimians and New World monkeys not falling into group 1
Group 3. Old World monkeys, great apes and humans

Napier also included the lesser apes in his Group 3. Nevertheless, it has been shown since then that the thumb of the lesser apes is not only unique in its mode of attachment, morphology, and proportions but also in its function. Human infants have been observed (by the author) using their thumb as a probe-finger and exploring textures and shapes in a manner very similar to the use of the thumb in lesser apes as reported by Lorenz (1971). According to Altner (1968), only *Tupaia* and *Tarsius* have truly hinge-shaped carpo-metacarpal articulations of their pollex. However, in *Tarsius* the main place of mobility of the thumb is not in the carpo-metacarpal joint—immobile in tarsiers—but between metacarpal and basal phalanx (Napier, 1961). In this respect tarsiers differ in manner of pollex movement from all other primates. However, other primates also have some mobility in the metacarpo-phalangeal articulation. All the primates that are grouped together as having pseudo-opposable thumbs do show morphologically variable degrees of incipient modified hinge or even shallow saddle articulations in the carpo-metacarpal region and thus are not a homogeneous group.

Functionally the hands of tree shrews and marmosets are very similar: both usually grasp with an adducted thumb and thus, all five clawed digits move in unison. It appears that this is so, in spite of the fact that some of the marmosets (e.g. *Callithrix jacchus*) show a proximal thumb morphology similar to a saddle articulation.

Most prosimians have thumbs that can be spread away from digits 2 through 5 to a considerable degree. This spreading capacity is especially pronounced among Lorisidae (Figure 6–20).

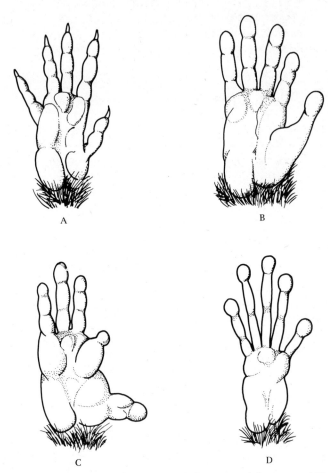

Figure 6–20 Volar aspect of prosimian hands: (A) *Tupaia*; (B) *Lemur*; (C) *Perodicticus*; and (D) *Tarsius*.

Most cebids show a tendency to grasp small branches and objects between second and third fingers. This type of grasping is reasonably typical for all the large South American monkeys and can frequently be seen as well in human infants. Species of the genera *Ateles* and *Brachyteles*, whose pollex is totally reduced, have to grasp between digits 2 and 3.

Only Old World monkeys, great apes, and humans are truly capable of opposability: rotating the pollex into opposition to the other digits in a way that puts the volar surface of the thumb onto the volar surface of the other digits. In correlation with the intricate nervous (tactile) supply of the volar skin, opposability makes skillful manipulation of even very small objects possible. However, even a few Old World monkeys have greatly reduced thumbs, as we shall soon see. In consequence, the monkeys have little ability to oppose their fingers 2 through 5. Among Lorisidae we find that the second finger is very much shortened (Figure 6–20). The thumb diverges at an angle of about 180 degrees away from finger 3. This gap with thumb on one side and fingers 3 to 5 together opposed forms a perfect pair of pliers. This arrangement makes them capable of an incredibly strong grip. Together with this, Lorisidae have a specialized vascular supply within their hands and feet: the so-called miracle nets (*retia mirabilia*) of capillaries. This sustained blood supply enables these animals to grasp a branch and hang on to it completely immobilized for hours, without hands and feet turning numb.

Humans, for example, cannot sustain a strong grasp around a dowel or handle for long periods, just because the blood circulation of the hands is not augmented by such a net of capillaries.

In tarsiers the volar finger pads of fingertips are enlarged and shaped like discs that can function as suction cups. Among tarsiers we also find that fingers are comparatively long, and metacarpals short.

Not only is the thumb reduced to a hardly visible stump in *Ateles* and *Brachyteles*, but also the same has happened among members of the Old World family Colobinae. Very short thumbs are the rule in genus *Presbytis*. Among apes, the orangutan has the shortest thumb. Macaques and baboons when walking place the volar aspects of the fingers 2 through 5 on the ground, whereas, in contrast, the African apes place the dorsal side of the middle and terminal phalanges of fingers 2 through 5 on the ground (Figure 6–21). This latter mode of locomotion, commonly referred to as "knuckle-walking," has been described and analyzed by Tuttle (1969).

In the gorilla we even find a friction skin on those dorsal aspects of the fingers that contact the ground during knuckle-walking. The positioning of the hands of macaques and baboons, as well as those of the African great apes, results in a slight elongation of the entire forearm; because with all these monkeys the length of metacarpus and carpus (and in the African great apes also the length of the basal phalanges) is added to the arm length when walking on their fingers rather than flat on their hands.

Another interesting primate specialization is that in the aye-aye all digits are clawed. The hand of the aye-aye is also highly distinctive: among the very long digits 2 through 5 (digit 4 is the

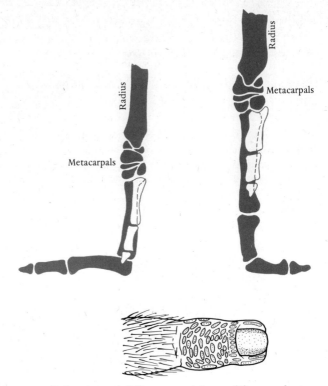

Figure 6–21 Schematized drawing of different positions of fingers during walking in macaques and baboons (left) and knuckle-walking apes (right) with the dorsal aspect of a *Gorilla* finger, showing the unique fingerprints. (Second finger = black, pollex = white.) (Dorsal aspect of finger redrawn after Biegert, 1961.)

longest), digit 3 is surprisingly thin and wirelike. At the end digit 3 is adorned with a large, curved, and pointed claw. This finger 3 is used to pull insect larvae, especially grubs, out from under the bark and cracks of trees and other hiding places (Figure 6–22). Claws are also found on all terminal digits of the hands of the South American Callitrichidae. These "claws" are regarded by some as secondarily reshaped nails. However, a layer of deep stratum—characteristic for claws even though it is very thin—is found in the claws of Callitrichidae. The German common name for the Callitrichidae even refers to the fact that these little monkeys have claws on fingers and toes except the great toe; they are called "Krallenäffchen," which means "litle clawed monkeys."

All other primates have nails on their hands. Among South American monkeys the nails are often less broad and flattened than

Figure 6–22 Skeleton of the left hand of *Daubentonia*.

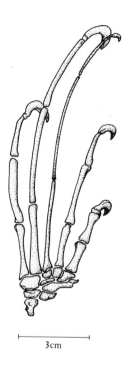

3cm

they are in many Old World monkeys; they are somewhat compressed laterally in many of the Cebidae.

On the inside of feet and hands (plantar and palmar surfaces), primates have cushioned pads that are covered by a very sensitive skin. The papillae of this so-called "friction skin" that covers soles and palms are arranged in parallel curved lines that form elevated ridges. Also, the ducts of abundant sweat glands of the skin open on the top of these ridges. This "friction skin" is also supplied with numerous nerve endings (Meissner's corpuscles) and thus is a close-up sense organ. The ridges of the friction skin occur in complicated patterns that are individually different in all primates and thus can be used for various identification processes.

One of the major mechanical or functional aspects of this friction skin is to allow a secure grip. In *Tupaia* both the first digit and the fifth digit can be spread away from the adjoining fingers to some degree. Altner (1968) conducted a very thorough study of the early ontogenetic development of the hand skeleton, focusing mainly on *Tupaia* as a "spreading hand." Interestingly, it turns out that the articulation between basal carpal and metacarpal of the pollex has the morphological structure of a saddle-articulation (movements possible along two axes that are positioned at right angles toward each other) in the mouse lemur. This same bone has the shape of a hinge-articulation (movement around only one axis) in *Tupaia glis* and also *Tarsius bancanus*. However, the saddle-articulation in the mouse lemur is rather shallow—and remains so in the adult animal. Therefore, it is regarded as an incipient saddle-articulation. It appears that, in addition to *Microcebus*, different grades of elaboration of the saddle can be found in *Lemur mongoz* and *L. macaco*. In *Galago senegalensis* and *G. crassicaudatus* this articulation is even more truly saddle-shaped. The articulation is deeper in both its convexity and concavity than in the hand of mouse lemurs or that of other lemurs mentioned above.

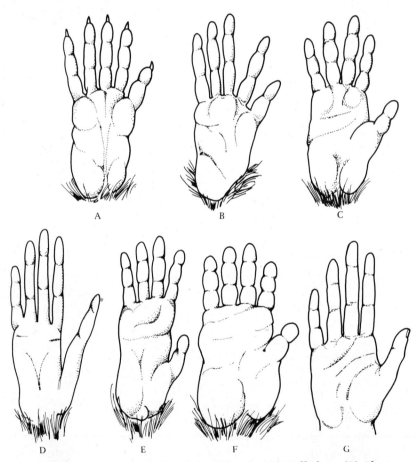

Figure 6–23 Volar aspect of anthropid hands: (A) *Callithrix*; (B) *Alouatta*, digits two and three spread, indicating the ways cebid monkeys grasp; (C) *Macaca*; (D) *Hylobates*; (E) *Pongo*; (F) *Gorilla*; and (G) *Homo*.

Some of the New World monkeys (*Callithrix jacchus, Leontocebus rosalia,* and *Cebus apella*) have hinge-shaped carpometacarpal articulations. However, the ceboids are of a type that appear to show slight morphological adaptations toward a two-axial saddle-articulation. In contrast, *Saimiri* has a saddle-shaped carpometacarpal articulation. However, these ceboids have a less perfect saddle-articulation than can be found in Old World monkeys, great apes, and man. The lesser apes are exceptional with regard to the carpo-metacarpal articulation of their pollex: it is ball-shaped and allows for an even wider range of movement of the pollex than does a saddle-based pollex. Lorenz (1971) showed that the gibbon's involvement of the pollex in various activities other than locomotor is exceptionally skillful (e.g. exploration of environment, of food items, grooming, and so on). The pollex also is comparatively long in lesser apes and separated from the index finger by a deep cleft that extends into the metacarpal region (Figure 6–23).

In most primates the hallux (big toe) is often large, but it is not opposed to digits 2 through 5 in the same way as is the case in primate hands. Except for Hominidae we can speak of a pseudo-opposition of the hallux in primates. Most primates have grasping feet. Only in one of the great apes, the orangutan, is the big toe reduced in length. The longest big toes, or great toes, are found among members of Lemuridae, Indriidae, and Lorisidae; see Figure 6–20. Short "big toes" are also found in callitrichids and terrestrial cercopithecids as well as in the orangutan. Relative to foot size, the big toe in man is not especially large. The toe appears to be big because the other toes are relatively short and small.

With some prosimians we find an elongation of the tarsal elements (calcaneus and navicular). This is true for Cheirogaleinae but is even more pronounced in Galagidae and Tarsiidae (Figure 6–24). Galagidae and Tarsiidae are very adept leapers, and the elongation of the metatarsals can be understood in this context. Tarsiidae show yet another specialization that can be interpreted as being related to the highly specialized locomotion of these animals: namely, tibia and fibula in the lower leg are fused together and elongated.

Among Lorisidae we find that the big toe diverges from the toes 2–5 in the angle that is close to 180 degrees (Figure 6–25).

Unlike the second finger, the second toe is not reduced among Lorisidae and it is adorned—as in all other prosimians—with an elongated claw. This claw is often referred to as the "toilet claw." The claw, indeed, is used in toilet activities such as grooming and scratching Among Tarsiidae we find two toilet claws on both the

Figure 6–24 Comparison of calcaneus (gray) and navicular (gray) elongation in: *Microcebus murinus* (left), an animal that does not habitually leap and cling; *Galago crassicaudatus* (middle), a large vertical clinger and leaper; and *Tarsius* (right), a highly specialized vertical clinger and leaper of comparatively small bodysize. (Brought to approximately same talus length.)

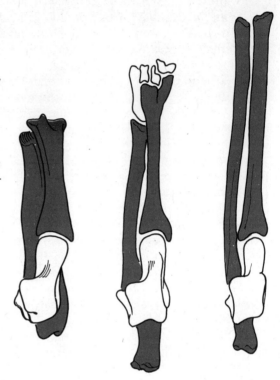

second and third toe as well. More primitive than primates, Tupaiidae have claws on all five toes, whereas all prosimians have a flat broad true-nail on their big toes. Thus, for example, the aye-aye has claws on toes 2–5 and a flat nail on the end phalanx of the hallux, as is also the case among South American callitrichids. The remaining New and Old World monkeys, the lesser and greater apes, and humans have nails on all toes.

Figure 6–25 Left foot of *Arctocebus calabarensis*.

Seven bony elements constitute the tarsus of the foot. the talus—also called *astragalus*—is the one element that articulates with the two long bones of the lower leg, tibia and fibula. The talus is covered by articular facets at its upper, lateral, and medial aspects. The lower ends of the tibia and fibula surround the talus from above and both sides, medially with the tibial malleolus and laterally with the malleolus of the fibula. Movements at this joint are mainly flexion and extension. The range of these movement varies between different species of primates. Medio-distally a robust process, the talar neck projects forward and slightly downward and terminates in a rounded thickening, the talar head. This talar head is covered by an articular facet and articulates distally with the navicular. The underside of the head articulates with the sustentaculum tali on the medial side of the calcaneus. The latter is the largest bony element of the foot, which in turn articulates distally with the cuboid; see Figure 6–18. In most simian primates the talo-navicular and calcaneo-cuboid or midtarsal joints are positioned adjacent to each other. They are, however, widely separated in such forms that have an elongated calcaneus. It is the navicular that elongates concomitantly with the elongation of the calcaneus (e.g. *Tarsius* and *Galago*, both highly specialized in their locomotion).

The navicular—as its name implies—is a boat-shaped element in most primates. The navicular has a concave joint proximally that embraces the talar head and a convex joint distally for the complimentarily concave surfaces of the distally adjoining three cuneiform elements. If the navicular is elongated, it becomes more or less quadrangular or, in extreme cases, long and cylindrical, as in tarsiers and galagos that show the most elongated navicular (and calcaneal) elements. The three cuneiform elements and the cuboid are joined by the metatarsals distally. The metatarsals articulate with the phalanges that usually have two elements for the first ray or hallux and three elements for toes 2–5. All these elements vary in relative length and shape between the different primate species. For example, an elongation of the metatarsals is found in many primates, such as the baboon (*Papio*) as well as in other mammals as tree-shrews (*Tupaia*), kangaroos, and many rodents. However, in all primates the general combination of the tarsal elements is fundamentally the same.

Great apes usually walk—when on the ground—with the weight on the outside of their feet. The feet of all non-human primates are efficient grasping tools. This condition is especially true for the feet of the orangutan that have the relatively longest toes (2–5) of all the higher primates, and yet the shortest hallux.

The human foot is clearly specialized: it is stiff and no longer good at grasping. Calcaneus and talus are positioned more or less at right angles to each other and upright to the ground surface. The neck and head of talus in *Homo sapiens* do not diverge laterally from the long axis of the foot as much as in other primates. This is presumed to be a consolidation for weight bearing and a morphological expression of the fact that the hallux has lost its grasping function in man. It is adducted, and thus, incorporated into the arch of the human foot. Among all other primates the hallux is a grasping toe.

In man the distal ends of all metatarsals (1–5) touch the ground and the proximal ends of metatarsals 1–4 do not: they are elevated into the arch of the metatarsus. Only the metatarsal 5 habitually touches the ground along its entire length at the outside of the human foot. The human foot is stiffened by strong ligaments to form a two-fold arch: arching both lengthwise and across the foot.

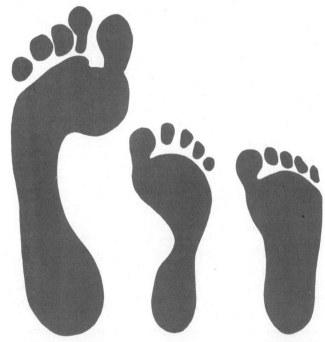

Figure 6–26 *Homo* foot prints of an adult (left), of a five-year old (middle), and of a two-and-a-half-year old (right), demonstrating the development of the arch of the human foot. The arch is missing at right, nearly complete in the middle, and fully formed at left.

This arch can be understood by looking at the footprint produced by a normal human adult. Such a footprint shows the arch clearly, for most of the central arched region does not touch the ground. The arch of the human foot only develops gradually during ontogeny. The arch is not found in children until about three years of age, when it slowly begins to rise and is nearly completed in children by about six years of age.

Considering the enormous length of the lever arm "body" in comparison to the length of the foot, the human foot in its entirety is comparatively small. Our durable foot structure has to support the upright body during standing, walking, running, dancing, and numerous other activities.

FIFTH EXTREMITY

Some South American monkeys have a versatile fifth extremity: their tail. On the ventral aspect of the distal third of the tail, they exhibit a friction skin just like that on their hands and feet. The tail is supplied with nerve-receptors and, thus, is highly sensitive. "Finger-print" patterns on the tail vary between genera, species, and individuals. Species of *Ateles, Brachyteles, Lagothrix,* and *Alouatta* of the family Cebidae all have this very "handy" fifth extremity. All these species are able to hang upside down by the very tip of their tails. These species can even play and hold another monkey in this position, so that both are solely suspended by the one tail. Scientists call this kind of tail a prehensive tail.

A functional and morphological incipient fifth extremity is found in genus *Cebus*. However, this tail is haired all over and thus not quite as useful as the true prehensile tails of the genera already mentioned (Ankel, 1972). That ancestors of apes and man long ago lost this most useful distal appendage, instead of developing an extra "hand," seems almost regrettable at times. Think of all the practical uses it could have had for us, and the great opportunities it would have offered the imagination of fashion designers.

Tarsius bancanus has a friction pad at the proximal ventral aspect of its tail that is used as a "tripod" during vertical clinging (Sprankel, 1965).

Many more details of comparative primate anatomy can be found in Swindler and Wood (1977) and other major publications on the anatomy of several primate species.

chapter 7

Sense Organs and Viscera

NOSE AND OLFACTION

In most mammals we find a moist and shiny glandular area around the nostrils; this area is the so-called rhinarium. Commonly, the rhinarium, an outward extension of the olfactory skin that covers the nasal passages, contains nerve receptors for smell and touch. At the middle the rhinarium runs down to the upper lip in most mammals (Figure 7–1). Besides housing the receptors of smell, the rhinarium also functions as a tactile organ for close-up perception. Inside and medially, the upper lip is attached to the gums or the upper jaw by a fold or mucous membrane, termed the philtrum. Such a philtrum is found among members of the families Tupaiidae, Lemuridae, and Indriidae. These primates with philtrum and rhinarium have also been classified together as Strepsirhini (Geoffroy, 1812) because of the structural similarity that they share in the nasal area. *Strepho* means "turned inward" in Greek, and *rhinos* is Greek for "nose."

In some of the bush babies and lorises, the rhinarium does not extend to the rim of the upper lip, but its labial part is also folded down into a deep medial groove and is entirely covered with hair. The naked area of the rhinarium in *Galago senegalensis* and *Microcebus murinus* even has a well-developed pattern of epidermic folds—analogous to the ridges on the palms of hands and feet—situated below and between the nostrils (Figure 7–2). Little wart-like structures are found on the rhinarium of *Lemur, Nycticebus,*

Sense Organs and Viscera

Figure 7–1 Prosimian rhinarium of *Perodicticus potto*. [Friderun Ankel-Simons photo; courtesy of the Duke University Primate Center.]

and *Perodicticus* and incipient tactile ridges are found in *Phaner*. Biegert (1961) has interpreted these tactile ridges on the noses of *Galago senegalensis* and *Microcebus murinus* as a highly developed close-up sense organ.

In the Tarsiidae and all anthropid primates, a rhinarium is

Figure 7–2 Rhinarium of *Microcebus murinus* with "finger prints" or, better, "nose prints."

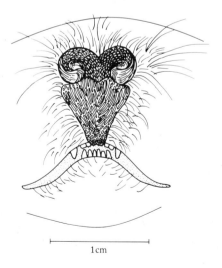

lacking or is restricted and dry. The upper lip is covered with hair, and the medial attachment to the gums in the premaxillar region has been totally reduced. Thus, the upper lip is freely mobile and is supplied with musculature that participates in changes of facial expressions that Strepsirhini are incapable of. This increased mobility of the upper lip plays an important role in social interactions among higher primates, and, last but not least, it allows humans to whistle. The dry rhinarium, common to one prosimian primate and all the Anthropoidea, has led some scholars to classify *Tarsius* and Anthropoidea together, and to contrast them to Strepsirhini as a separate group Haplorhini (Pocock, 1918). *Haplo* means "simple" in Greek. This classification, however, is not widely accepted because there are many valid reasons to classify the aberrant prosimian *Tarsius* separately from Anthropoidea, but not necessarily with Strepsirhini. The moist glandular skin of the rhinarium, that has been lost among the tarsiers and the higher primates, has been thought to have been a loss because of an assumed weakened olfactory ability among all such primates. Also, the nasal passages have other uses besides air ducts: They have warming and cleaning functions and provide olfactory information at the same time.

It once was commonly assumed that the sense of smell has little importance for primates. Even the foreshortening of the snout in higher primates has frequently been correlated with an assumed reduction in the sense of smell among the larger and more advanced Anthropoidea. Although such reduction is true if we compare primates with animals like the dog, which relies to a great extent on olfaction, higher primates do retain a good sense of smell, and it plays some part in their life. Why else would New World monkeys have scent glands? Why would they engage in urine washing? It seems less surprising that among prosimians urine-marking of territories (*Perodicticus potto, Microcebus, Cheirogaleus*) happens. *Lemur, Hapalemur,* and *Varecia* also scent mark with scent glands, located on their wrists or elsewhere. All these prosimians do have comparatively large snouts and nasal areas. However, the only measure of an animal's capacity in smelling is the size of epithelium containing smell receptors and the density and number of these receptor cells. Even in man the sense of smell has some importance. However, for reasons that are easy to understand, there have been no studies undertaken that would clarify the absolute degree of olfactory ability among the whole range of primates. The animals would have to be asked to calibrate the degrees of perception! Areas of body skin that are covered with glands have been described in detail for many South American monkeys (Epple and Lorenz, 1967).

These glandular areas are used in scent marking, an activity that requires that the animal species doing it be capable of a high degree of olfactory perception. As already stated, if we compare primates with relatively large bodied terrestrial animals like dog or wolf, the sense of smell seems less important among primates than in Canidae. However, it is known that at least among New World monkeys scent marking plays a considerable role within the framework of sexual, aggressive, and territorial behavior. One group, working in South America, reported that the proximity of *Alouatta seniculus* (howler monkeys) was noted because of their specific odor that could be detected before the animals were actually seen (Collias and Southwick, 1952).

As mentioned before, the very nature of the sense of smell makes an exact study of the olfactory capabilities of primates extraordinarily difficult, and consequently, no reliable data are available.

The decrease of importance of olfaction and taste, beginning from a Cretaceous "smell-oriented" ancestor, has often been assumed to have correlated with increasingly more and more advanced levels of organization within the Order Primates.

Recent studies have been conducted by Glaser (1972a) that now make it possible to compare the tasting abilities of primates with those of nonprimates. The traditional generalization about the trend toward reduction of olfaction and taste among primates has also been based on the superficial observation that the number of nasal turbinal bones and associated conchae is reduced in primates. Then, this reduction has been equalled with an assumed concomitant reduction of the level of taste and smell abilities. As already stated, only the absolute number of receptors of the olfactory nerve (*bulbus olfactorius*) per surface area of the nasal mucosa are relevant for assessment of the olfactory abilities of the particular animal species. The *nervus olfactorius*—or olfactory nerve—is usually regarded as a part of the brain itself and not as a peripheral nervous ganglion. The surface area of the nasal mucosa proper does not necessarily give information about the olfactory ability of an animal. Napier and Napier (1967) distinguished between primates with an "olfactory" muzzle, saying that it is characteristic for "certain Lemuridae and Lorisidae," and a "dental" muzzle, characteristic for certain ground living Cercopithecidae and for Pongidae because of their enlarged masticatory apparatus. Among Cebidae a similar deepened muzzle is caused by the enlargement of the larynx in *Alouatta*, the howler monkey.

Considering noses, we can state that prominent protrusions

like those of man are not very common in other primates. Three species of colobines that live in remote areas of Asia, however, do exhibit prominent noses. Two primate generic names for these colobines have been descriptively based on their possession of impressively large noses: *Nasalis* and *Rhinopithecus*, derived from *nasus* in Latin and *rhinos* in Greek, both words for "nose."

Genus *Nasalis*—also commonly known as "proboscis monkey"—has the largest nose to be found of all the primates, for the male proboscis monkey even outdoes humans in this respect. In the adult male *Nasalis* the rather bulbous nose hangs down past the mouth and even beyond the chin. In females the nose is considerably smaller and slightly turned up. Whenever the male proboscis monkey is aroused, his nose may swell and turn reddish. In Southern China and Western and Northern Vietnam, genus *Rhinopithecus* survives in relatively remote areas. Both sexes have snub noses that are not very big. The third "nosed" genus is *Simias*, a medium-sized monkey that only inhabits some small islands off the coast of Sumatra. Simias monkeys are snubnosed like *Rhinopithecus*, but their noses, although prominent, are still smaller than in the latter. All these enlarged noses have been interpreted as secondary sex characteristics. Additionally, in male *Nasalis*, the nose seems to function as an organ of resonance in vocalization, rather like a horn. An occasional individual with something like a nose can also be found among members of species of genus *Colobus*. From this we can conclude that prominent noses among monkeys are restricted to the Old World colobines.

The nasal region of the face of gorillas is also especially interesting (Hofer, 1972). Not only is the flat nasal area large in relation to the size of the face, but also it can be quite conspicuous. The morphology of the nose in gorillas varies widely intraspecifically and thus can easily be used to identify individual animals. In field studies this characteristic difference of gorilla nasal areas has been proven very useful for identification (Schaller, 1963). Another feature of the nostrils in primates is of interest: in prosimians the nostrils are shaped by cartilage where the nose extends beyond the bony nasal skeleton. In lemurs cartilaginous tubes extend into the tip of the nose and are positioned more or less parallel and close to each other. These tubes diverge in tarsiers and consequently, the nostrils open sideways in these animals. Old and New World monkeys can easily be distinguished by the way in which the nostrils are positioned. The nasal openings in Old World higher primates are positioned close to each other and directed forward or downward, whereas in the New World monkeys the nostrils are far apart from each other. These differences are caused by the positioning of the cartilaginous wings that support the nostrils. Among higher primates the nostrils are only tubelike early during ontogenetic development. Solely in the Old World Anthropoidea including man, a differentiation into separate cartilaginous wings takes place. In New World monkeys only a lateral split appears in the nasal cartilage, but it does not lead to separation into independent cartilages. The clefted cartilaginous tubes open sideways in New World monkeys and thus cause the relative greater distance between the nasal openings in these animals (Wen, 1930). This difference can be easily perceived in the faces of all monkeys (Figure 7–3) and apes and leads to the taxonomic separation of New and Old World Anthropoidea into Platyrrhini and Catarrhini respectively (Hemprich, 1820). The validity of this "nose-classification," however, has also been widely discussed and argued against, just as the terms Strepsirhini and Haplorhini have been criticized. Unlike the latter pair of terms, however, the terms Platyrrhini and Catarrhini have been kept current ever since they were coined and are a useful distinction between the higher primates of the New and Old Worlds that is maintained in this book. Even if one dislikes the distinction of Platyrrhini and Catarrhini for taxonomic reasons, these terms are certainly surviving. It seems that no harm is done by

Figure 7–3 Broad—Platyrrhine—and small—Catarrhine—nasal septa. (Left) *Lagothrix* [photo courtesy of Michael D. Stuart]; (right) *Pongo* [photo courtesy of Dieter Glaser].

this taxonomy either, unlike the use of the terms Strepsirhini and Haplorhini that give neither colloquial nor taxonomic precision. Tarsiers, in grade of evolution, do not belong with Anthropoidea—the two disparate groups are covered under the term Haplorhini mainly on account of this one single nasal characteristic that is similar in both. Platyrrhini and Catarrhini, in contrast to this, are two large and varied groups of primates that are taxonomically compatible because they are quite homogeneous in evolutionary grade and structure.

ORAL CAVITY, TONGUE, AND TASTE

The top of the oral cavity of primates is characterized by differently structured palatal cross-ridges (*Rugae palatinae*). These palatine ridges—structures of the hard palate and the mucous membrane—are found in all mammals except in the odontoceti (toothed whales). Such ridges are of mechanical help in holding and transporting food items within the oral cavity and even aid in the processing of food by chopping it. Palatine ridges appear early in ontogeny and do not change their number or shape later (Schultz, 1958). Rugae are well-developed in all prosimian primates, usually occurring in numbers as low as five to six among Lorisidae. In *Tarsius*, even though this animal has a very short snout, rugae may be as many as ten to eleven. It is typical of prosimians that these ridges cover the hard palate back to the end of the last molars. In higher primates the ridges usually do not extend as far backwards and also are often of irregular patterns unlike those in prosimians. The number of rugae varies from around four to eight among Anthropoidea.

Within Hominoidea these ridges are perhaps the most reduced and irregular. In man, usually only four comparatively shallow ridges cover the palate between premolars and canines.

Tongues have a number of functions. Besides helping with intake and positioning of the food items within the mouth, this structure participates in vocalization, and its main function is testing and tasting food and drink. The tongue is also involved in cleaning the mouth and teeth.

The principal taste receptors of primates are located on the tongue. On its surface the tongue is covered with numerous papillae that give it a rough surface. This surface structure is important for the transport and sensory exploration of food items. Normally, the papillae can be divided into four types that are shaped differently and also have different functions. The majority of the papillae are small and usually end in a few points or in filamentous processes that give these papillae their name. Such filiform papillae have a brushlike texture, and thus are well equipped for transportation and holding of food items. Less numerous and larger are the fungiform papillae, *fungus* being the Latin for "mushroom"; these are named because they are shaped like mushrooms, having a stalk expanding toward their tops. A third kind of papillae that are called simple papillae are distributed all over the entire mucous surface of the tongue. Each fungiform papilla contains a capillary loop. The fungiform papillae are concentrated in their occurrence at the tip of the tongue and on its sides; they contain taste buds. However, the majority of taste buds are located on the largest type of papilla, the vallate papillae. The term vallate papillae describes the shape of those papillae that are structured like a truncated cone. The largest diameter of the cone points away from the tongue, and the smallest diameter is at the attachment of such cones. These buds are surrounded by a circular wall. Papilla and wall are separated by a furrow or duplication of the epithelium. Especially on the wall and within the furrow we find a high concentration of taste buds. The vallate papillae are located on top of the tongue at its base; the number of these is small and varies between eight and twelve in modern man. Taste buds are also scattered over the sides and back of the tongue and adjacent parts of the mouth.

With primates of small body size, one usually finds only three of the vallate papillae on the base of the tongue; there they are arranged in a triangle. It is said that the number of vallate papillae varies according to absolute size of the tongue.

Tongues of the members of the infraorders Lemuriformes and Lorisiformes are different from those of other primates because the tongues have a dense field of cornified papillae (these being somewhat larger and longer papillae of the filiform type) beyond the vallate papillae down towards the throat. These tongues have clearly developed mechanical functions to facilitate swallowing.

A duplication of the muscular primary tongue is found below the tongue itself and therefore is called the sublingua or undertongue. This structure appears in many primitive mammals and also in prosimians in which, according to Wiedersheim (1902), it seems to have the most pronounced development. Structures that are sometimes found on the under surface of the tongues in monkeys, apes, and men may or may not be remnants of the prosimian sublingua.

A sublingual organ was described in *Callicebus* by Hofer (1969).

According to Schneider (1958), it is, however, uncertain whether any of the sublingual structures of higher primates and other mammals could be considered homologous with the sublingua of Prosimii. Among prosimians the sublingua is well developed and cornified; its medial axis is a thick structure that attaches the sublingua to the under surface of the tongue. Only the tip of the sublingua is free and mobile. The sublingua extends below the tip of the tongue in Lemuridae as well as among galagos and lorises; its tip is cornified and splits up in several serrated points. The undertongue's function has been described by Bluntschli (1938) as a "toothbrush" for the front dentition in those forms that, as we know from the teeth, have a specialized procumbent tooth comb. As a matter of fact, the median thickening of the sublingua is equipped with hook-shaped structures in *Daubentonia*, the Madagascar aye-aye. These hooks are obviously a specialization that correlates with the very unusual front dentition of *Daubentonia*. The hook-shaped sublingual tip fits perfectly into the interspace between the two lower incisors and thus keeps this area clean.

In *Tarsius* the sublingua is shaped more simply and lacks the serration of the lemurs. This difference is again correlated with the fact that tarsiers do not have the specialized lower tooth comb of the other prosimians associated with a serrated undertongue. The surface of the tongue of *Tarsius* is different from that of all the other prosimians in having the tips of the filiform papillae split up into numerous filaments. Also, the cornified papillae on the upper surface of the root of the tongue are missing in *Tarsius*.

Among marmosets or Callitrichidae we do find a small sublingua that more or less resembles the sublingua of *Tarsius*. This sublingua is only weakly serrated and does not extend as far as the tip of the tongue. A fold of mucous membrane skin that is found along the edges of the lower surface to the tongue in man, the apes, and Cebidae is called the *plica fimbriata* (*plica* is Latin for "fold" and *fimbriatus* the expression for filamented). This fold is often thought to be a remnant of the sublingua of marmosets and prosimians. The same structure is even more reduced or missing among the various cercopithecoid monkeys.

Hofer (1969) studied and discussed some sublingual structures in the South American monkey *Callicebus* and later in *Perodicticus potto* (Hofer, 1971). He states that the structure that has been described as sublingua or as frenal lamella in *Callicebus* is not a real sublingua. He found that this structure—unlike anything described for other primates—contains the excretory ducts of salivary glands around which are located taste buds. The body of the organ lacks any musculature and therefore moves passively together with the

Sense Organs and Viscera

tongue. Also, among prosimians the true sublingua does not have its own active musculature either but moves passively together with the tongue; see Schneider (1958). Ever since Bluntschli (1938) first introduced the functional interpretation of the sublingua in prosimians as a toothbrush, there has been little or no clear confirmation of this function. Even now we lack careful anatomical and functional studies of the sublingual organs of most primates or among insectivores and marsupials; studies that would allow for more precise definition of the function and origin of this structure among prosimians.

As we have seen, there are not any reliable data on the sense of smell among primates for reasons that are easy to understand. This condition, however, is not the case with the sense of taste. Taste experiments are easier to conduct than are assessments of smell. Other than in a few higher primates, nothing much was known until about ten years ago about the tasting abilities of most primates for all the qualities of taste, namely: sweet, bitter, sour, and salty (Figure 7–4). Because in mammals—including all primates, taste receptors are located within the mouth and predominantly on the

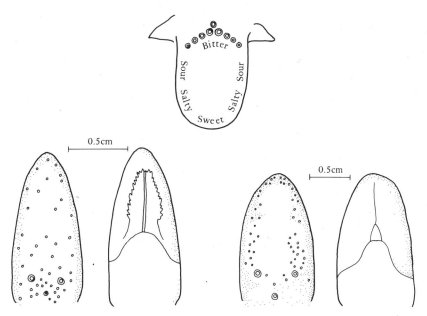

Figure 7–4 (Top) tongue indicating the places of taste for the different taste qualities. (Left) prosimian tongue, upper side and under side showing serrated sublingua, and (right) tongue of a marmoset, upper side and under side with small, reduced sublingual structure. (Double circles = circumvallate papillae.)

surface of the tongue, the quality and—also more importantly—the quantity of taste-inducing substances can be exactly measured. In 1968 Glaser of Zurich published the first results of an ongoing series of taste tests with various primates. His thorough studies have furthered our knowledge about the tasting abilities in primates and man. The results show that in most of the components of taste, human sensitivities are equal to or even better than those of the other primates tested. It also appears that primates in general have about the same abilities in taste perception as do other mammals that have been tested such as rodents, lagomorphs, artiodactyls, carnivores, edentata, and even marsupials. It seems that in the four taste qualities, there is no difference in ability to recognize tastes between prosimians and Anthropoidea (Glaser, 1972a). It also appears that man has the ability to detect bitterness (chininhydrochloride) and sourness (acetic acid) in a much lower concentration than the other primates that have been tested. In general, it appears that only rarely are man's tasting abilities inferior to other mammals that have been studied. Hellekant et al. (1976) have sought for the possible mechanism of gustatory effects, correlating behavioral reactions with electrophysiological responses of the chorda tympani nerve proper that mediates tastes sourness, saltiness, and sweetness from the anterior part of the tongue. The taste quality "bitter" is perceived at the base of the tongue and mediated by the glossopharyngeal nerve. A few tastebuds are thought to be located in the epiglottis, and these are supplied by the vagus nerve.

Interestingly, Glaser (1972b) has also measured variations in tasting ability of man of the chemical phenylthiocarbamid commonly known as PTC. He was able to show that the ability to taste (or not to taste) PTC cannot be used to distinguish different races or populations of humans, as many anthropologists had believed.

Recently, Glaser et al. (1978) have tested a wide array of primates and some representatives of six major other mammal families concerning their response to two proteins: thaumatin and monellin. Both these proteins are known to elicit an intensely sweet taste sensation in man. Interestingly, all nonhuman catarrhine primates—Cercopithecoidea and Hominoidea—show the same strong reaction to these two proteins electrophysiologically and behaviorally as does man. All the rest of the primates, as well as Tupaiidae and the remainder of mammals tested, do not react at all to thaumatin. A few exceptions for this are found with monellin: *Varecia variegata, Cebuella pygmaea, Saguinus fusciollis, S. nigrifrons*, and *Saimiri sciurea* slightly prefer monellin to water, and *Lemur mongoz* strongly prefers a solution of monellin to water.

Glaser is now continuing his study to find out what causes these differences of tasting ability between the catarrhine primates on one hand and the other primates and mammals on the other hand. Glaser et al. (1978) state that from a gustatory point of view, the Catarrhini might as well be called "Thaumatina."

AUDITORY REGION AND HEARING

The scientific term for the outer ear is the *auricle* or the *pinna;* it is formed of an elastic cartilage that is covered with skin and connected to the skull by ligaments, muscles, and fibrous tissue. The central part of the external ear is called the *concha,* a word derived from the Greek through Latin that means a "mussel shell." The concha is attached to the head at the external opening of the ear and forms an irregular "funnel" around it. The part of this funnel extending above the ear opening is called the *scapha.* The *scapha* together with an upward and backward extension forms the helix. If this reaches below the ear opening, it forms the lobule or ear lobe. The outer opening of the external auditory meatus is partly covered by two cartilaginous extensions, the *tragus* in front and the smaller *antitragus* in the back (Figure 7–5). A more or less pronounced transverse fold, the so-called *plica principalis,* delimits the concha in its upper part and is sometimes enlarged into a flap. This, for example, is the case in bush babies (*Galago*) and tarsiers (*Tarsius*). Thus, we find that in members of the two latter genera, the upper part of the auricle is very large and is equipped with additional crossfolds. These large membranous outer ears are not rolled in at the margin and can be folded down towards the ear

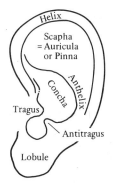

Figure 7–5 Ear.

opening with help of the intrinsic musculature (*musculus corrugator pinnae*) that is incorporated into the auricle. This muscle—where it is functional—changes the shape and direction of the outer ear and folds it down. Bush babies and tarsiers can move both ears independently from one another. The *musculus corrigator pinnae* is absent in all Anthropoidea, including man. The entire outer ear can be moved by several extrinsic small muscles (*m. depressor auris, m. retrahens auris, m. rotator auris, m. attrahens auris,* and *m. attolens auris*) that respectively pull the ear down, pull it back, rotate, pull forward and upward, and pull upward only.

Among anthropoids the outer ear consists mainly of a single piece of cartilage, covered with skin, that gives the auricle its shape. This cartilage does not extend into the lobule of the ear. The lobule is composed of areolar and adipose tissue and is rather soft and pliable, unlike that stiffened part of the ear held up by cartilage.

Where the outer ear is concerned, great variability between and within primate genera, even species can be observed. Thus, for example, Lasinsky (1960) was able to show that not only the shape, but also the relative size of ears is most variable throughout Order Primates. With prosimian primates differences in size and shape of the outer ear are extreme. These variations in size and shape of the outer ear are related to behavioral characteristics that call for different ear functions. Two functional factors influence the morphology of the auricle. One purpose is to serve as a funnel to collect sound waves, and another—if the auricle is large—is as a heat regulating surface (as, for example, in elephants: Hesse, 1920). A highly vasculated auricle with a large surface considerably increases the cooling area for the bloodstream in mammals that inhabit hot climates. The blood vessels in this large surface can be contracted or expanded, thus changing the amount of blood-flow and the cooling efficiency of the blood stream. An example of the function of the auricle as a megaphone can be seen in the difference in morphology and size of the auricle between the two closely related genera: *Ptilocercus* with large, mobile ears and *Tupaia* with small, restricted outer ears. The basic behavior of these two genera is also different: Pen-tailed tree shrews are mainly nocturnal animals while tupaias are diurnal and/or crepuscular animals. Consequently, the large ears in *Ptilocercus* would suggest that it is more dependent on hearing than are tree shrews, which can rely to a greater degree on their visual sense. However, no studies have as yet been undertaken to prove this inferred hearing difference between the two genera. By studying behavior in both groups, it ought to be possible to demonstrate a distinction in hearing abilities and also to see

Sense Organs and Viscera **213**

Figure 7-6 (Left) *Tupaia* ear; (right) the ear of *Cebus*, an anthropoid monkey.

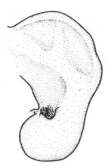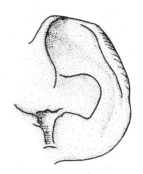

correlated separation of structure in the central nervous system. The outer ears of *Ptilocercus* are large and membranous, as in bush babies, and very mobile. The ears of *Tupaia* resemble the external ears of higher primates and even man in a striking way (Figure 7-6). They are neither membranous nor very mobile.

We also find distinct differences in the morphology of the outer ears of two closely related prosimian families—the Galagidae or bush babies and Lorisidae, the lorises and pottos. The former have large, membranous, mobile ears that can be folded in and down, and the latter have comparatively small ears that are not very mobile. Again, a striking difference of the general patterns of behavior is observed: galagos are nocturnal and are very active, excitable and agile animals, whereas lorises—also nocturnal—are slow climbing and deliberate in their movements. Here also, we do not have any published evidence for the theoretical explanation of these differences. No comparative behavioral, physiological studies, or comparisons of the nervous systems of these animals have been undertaken to prove the supposed difference in ear function. As we have seen, the outer ear of *Tupaia* is relatively small and strikingly similar in its appearance to the ears of anthropoid primates. The opening of the outer ear is overlapped by the *tragus* from the front. Starting at the *antitragus*, the *helix* extends backwards, bending forward and upward to form the *anthelix*. The ears of *Tupaia* are bare, roughly round in shape, and positioned close to the head. In the area of the *helix* the ear is rolled up.

The outer ears of most lemurs are simple in their morphology and frequently adorned with hairtufts or evenly covered with hair. Lemur ears usually stick out from the head; *tragus* and *antitragus* are simple; and the *anthelix* and *helix* are also of simple shape, with the *helix* not rolled in (Figure 7-7).

Species of genera *Microcebus*, *Phaner*, and *Galago* have large, membranous ears. Among each of these groups, this large ear shows

Figure 7–7 Simple prosimian ear of a young *Perodicticus potto*. [Friderun Ankel-Simons photo; courtesy of the Duke University Primate Center.]

a variable number of several small crossfolds (Figure 7–8) parallel to the *plica principalis* that play a functional role when the large membranous earflap is folded down by the action of the intrinsic *musculus corrigator pinnae*. In a sound-polluted world would it not be wonderful had we the ability to occasionally fold our ears down to cover the ear opening? The *supratragus* (a flap covering the ear opening from above) is present but comparatively small among these small lemurs and the bush babies. Also, in *Tarsius* we find a very large, membranous and mobile ear. Here, the *antitragus* flap is larger than in *Galago* or *Microcebus*. It is known that the hearing ability of *Tarsius* is very acute, and that the terminal nuclei of the auditory nerves and their connections with the central nervous system are especially large. Throughout the anthropoid primates we find that the ears are comparatively small and of the same overall pattern. Usually, the ears are rounded, and the margin of the *helix* is more or less rolled inward.

The upper part of the *helix* is not rolled in and is pointed in two genera of the Cercopithecidae, namely in *Macaca* and *Papio*, comparable to the pointed ears of many lower mammals. Anthropoid ears never go through a pointed stage during their ontogenetic development except in the two genera just mentioned.

Sense Organs and Viscera

Figure 7–8 Complicated, large prosimian ear with crossfolds of a *Microcebus murinus*. [Photo courtesy of Michael D. Stuart.]

Schultz (1965, 1969) states that the pointed ear of an "orangutan fetus" that was pictured and described by Darwin (1871) (Figure 7–9) was caused by a deformation of that particular fetus. Moreover, in Schultz's judgement, the fetus is that of a gibbon not of an orangutan. Also, in Darwin's "Expressions and Emotions in Man and Animal" we find two drawings of the head of a Celebes macaque that show a pointed ear (Figure 7–10). This point of the ear auricle has gone into natural history lore as the *"tuberculum darwini"* or "Darwin's point." In man, where the point is actually hard to find and is still regarded by many as an atavism. Many human anatomy texts compare the "auricular tubercle of Darwin" with the pointed ears of "adult monkeys." Lasinsky, however, shows that the two structures have nothing in common. The auricular tubercle of Darwin had a recent and rather exaggerated revival in the very pointed ears of Dr. Spock—one of the mythical people of "Vulcan" who evolved from the fantasy of the creators of *Star Trek*.

Figure 7–9 Darwin's fetus with pseudo-pointed ear.

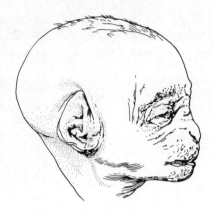

Among apes and man the rim of the aricle is usually rolled inward at the upper and the hind margin; See Figure 7–6. Among the large apes we find very small ears in orangutans, small ears in gorillas, and very large flaring ears in chimpanzees. Moreover, a high degree of variability in ear shape characterizes the common chimpanzee, *Pan troglodytes*.

An ear lobule is not only typically found at the lower end of the ear of modern man but also can be detected in the ears of African apes and some of the Old World monkeys, as, for example, in langurs. The lobule never appears in any of the prosimians, New World monkeys, lesser apes, or orangutans.

Figure 7–10 Darwin's point of the ear in a Celebes macacque. *Cynopithecus niger* in a placid condition and when pleased by being caressed.

In all Old World monkeys, apes, and man the inner ear and the eardrum are connected to the outside of the head by an ossified tube. In contrast, all New World monkeys have the tympanic membrane (eardrum) positioned much closer to the surface of the head, and the short acoustic meatus is made up of cartilage. With the tree shrews and the prosimians the membrane is also situated more closely to the outer ear opening than in Old World monkeys, apes, man, and tarsiers, as we shall soon see. The tympanic membrane is only suspended by a slender ring in Tupaiiformes and Lemuriformes. Among Tarsiiformes we find a long ossified acoustic meatus from the bulla to the outer ear opening that can be interpreted as a consequence of the highly specialized overall morphology of the entire head in this peculiar animal. In Lorisiformes the tympanic membrane is ossified to the skull (Werner, 1960). Acoustic vibrations are picked up by the membrane and transferred to the delicate series of three ossicles that are attached to the tympanic membrane, to each other, and to the skull by means of ligaments. All these bones are housed in the inner ear cavity (tympanic cavity). Ultimately, the sound waves that are transferred by the ear ossicles are picked up by the stato-acoustic nerve. The first of the ear ossicles, the *malleus*, is attached to the tympanic membrane; its name comes from its shape, similar to a hammer, called *malleus* in Latin. Adjoining the *malleus* is the *incus*; it has a rounded body and two thin, leglike extensions. The *incus* does not really look like an anvil, but that is the meaning of *incus*. The third ossicle is called *stapes* (Latin for "stirrup"), and it really does resemble a stirrup in its shape. This bone connects to the vestibule of the inner ear, or to be more precise, to the membrane that covers the so-called oval window, or *fenestra vestibuli*, that is situated in the lateral wall of the vestibulum of the inner ear. The inner ear or labyrinth is the main organ of hearing and also the organ of equilibrium, where the statoacoustic (vestibolocochlearis) nerve receives acoustic and equilibrial impulses.

There have been a few attempts to use the ear ossicles of primates for taxonomic purposes. However, nothing much is known about the variability of ear ossicles or the dependence of their shape on possible differences in function, facts that have been overlooked by the "ossicle taxonomists". Some general statements, however, can be made: Ear ossicles of Tupaiiformes are more like those of some lemurs than like those of other insectivores (Doran, 1876a,b). In some features, lemur ossicles resemble those of Old World monkeys. Interestingly, those of *Daubentonia* show some resemblances to earbones of certain rodents. Tarsier ossicles are similar to those of lemurs, whereas the ossicles of Lorisidae resemble those of Callitrichidae. Cebidae and Cercopithecidae each have earbones with some characteristics of their own, and the ear ossicles or Pongidae are quite similar to those of humans.

It has been established that only the chimpanzee and man among nineteen mammals that have been tested do not have the ability to hear high frequency sounds (Hefner et al., 1969a,b). Tree shrews and bush babies as well as marmosets are capable of perceiving very high frequency tones. *Homo* and *Pan* cannot hear sounds in high frequencies above 32 Kc/s (kilocycles per second) that can be perceived by bush babies and tree shrews. It also appears that bush babies are more sensitive to the discrimination of such frequencies than are hedgehogs and opossums. Man is especially capable of perceiving low-frequency sounds. Our overall sensitivity to sound discrimination is better than that of any other mammal so far tested (Hefner et al., 1969a,b).

EYE AND EYESIGHT

All primates and especially higher primates are visual animals. The structure of the primate eye appears to be related to activity patterns. The primate way of life is also reflected in the substructure of the area of light reception, the retina. On its backside the eyeball is enveloped by three different cell-layers, the innermost of which (towards the center of the eyeball) is called the retina. The retina is the area of reception of optical impulses and is also regarded as directly continuous with the optic nerve. The retina itself consists of an outer pigmented layer and an inner layer of nerve cells that, in turn, contact the vitreous body of the eye. The inner layer of the retina usually is made up of two types of nerve receptors. These receptors are cylindrical rods that are arranged perpendicular to the surface. Within the outer portions of these rods, visual purple or rhodopsin is found. Rhodopsin is a pigment that facilitates the absorption of low light intensity. Thus, rods are sensitive to low intensity light and black and white discrimination. Rods predominantly function at night. The other type of light receptors are called cones and are conical or flasklike in shape. The broad ends of these receptors are directed towards the center of the eye. The pigment in the outer portion of the cones is called iodopsin. Rhodopsin and iodopsin are both photosensitive receptor pigments. The cones function to give visual acuity and color reception. Animals that have predominantly rod retinae are usually night active (nocturnal) or with a few cones active during dawn and dusk (crepuscular), and those animals whose retinae are made up of a higher number of cones than rods are mostly active during daylight (diurnal). Among higher primates the retina is usually composed of both types of light receptors and also has a so-called macular area and a *fovea centralis*. The macular area (*macula lutea*) is a spot that is located in the optical axis of the retina and has an even smaller dimplelike depression in its center, the so-called *fovea centralis*. The macula consists only of cones that usually are more tightly packed in this area than anywhere else within the retina. The greatest concentration of cones occurs in the fovea, which is regarded as the place of greatest visual acuity.

Only the South American night monkey (*Aotus trivirgatus*) stands as an exception to the rule that higher primates have both cones and rods and that cones outnumber the rods. This monkey only has a few cones, if any, among the rods that cover its retina. *Aotus* is also reported to have a vestigial fovea centralis. Even though students of primate retinae do not agree on the question of, how many (if any) cones the *Aotus* retina contains (could this possibly be individually variable?), it is undisputed that the *Aotus* retina predominantly contains rods, and is very similar to the retinae of nocturnal prosimians and quite unlike those of other higher primates.

The retina of *Tupaia* is predominantly a cone retina (compare Rohen and Castenholz, 1967); that is, it is the retina of a diurnal animal. Tigges (1964) concluded from tests that attempted to check the ability of tree shrews to recognize colors—contrary to common belief—that they discriminate brightness rather than hue. The periphery of the retina of each eyeball is thicker in tree shrews than the inside of the retina. This fact has been brought up in connection with the belief that *Tupaia* (in spite of having the eyes situated more at the sides of the head than on the front) has already attained a certain degree of binocular vision (Wolin and Massopust, 1970). In *Tupaia* the arteries and veins that supply the retina show an arrangement that is quite different from that of any primate. The arteries and veins are located like the radii of a circle, protruding from the papilla of the optic nerve; usually they do not show such a definite pattern of distribution but supply the retina in a rather random netlike pattern. There is a nonvascular area in the retina of species of *Tupaia* studied so far. Also, the retina of *Urogale everetti* has been studied and found to be similar to that of *Tupaia*. However, the retina lacks the thickening at the outside of the eyeball seen in *Tupaia*.

The retina among Lorisidae and Galagidae has been studied in a number of genera. All retinae of these genera have been described as being of nocturnal type in basic accordance with their activity patterns. In many mammals such as carnivores and pinnipeds, an additional cell layer can be found within the choroid that envelops the retina and is, like the choroid, extensively vascularized. This highly reflective layer is called the *tapetum lucidum*, and it is the source of the brilliant reflection that makes eyes with such a tapetum light up in the dark. The tapetum is missing in *Tupaia* but has been found in the choroid of all the Lorisidae and Galagidae that have been studied. Few, but sparsely distributed and rudimentary cones, have been found in two lorises: *Nycticebus coucang* and *Loris tardigradus*. Among Lemuridae a great number of reports on retinal formation and eyeball morphology have been published. For example, the retina of *Lemur catta* has been described by several scholars. They all agree that, in spite of the fact that *L. catta* is predominantly diurnal, the retinae of the ring-tailed lemur seem to be similar to those of nocturnal prosimians. In ringtailed lemurs, the photoreceptors are mostly rods. Also, a tapetum is present. A behavioral study, however, shows that *L. catta* and *L. mongoz* are capable of some color discrimination (Bierens de Haan and Frima, 1930).

In these lemurs there is a well-defined *area centralis*, forming a dome like retinal thickening in the center of the retina. This area is not the *fovea centralis* that has been reported only in *Hapalemur* and *Lemur catta*. Most retinae of those lemur species that have been studied seem to be made up almost entirely of rods, although in *L. catta* a ratio of one cone to five rods has been reported. Only some species of *Lemur* appear to lack a tapetum. Both *Cheirogaleus* and *Microcebus* have retinae with only rods as photo receptors, and *Microcebus* does have the tapetum, said to be missing in *Cheirogaleus*. Also, *Avahi* is reported to have both tapetum and a rod retina thickened towards the center. *Propithecus verreauxi* and *Indri indri* have rod retinae with a few large cones scattered between them; they also are reported to have a distinct but comparatively small central area. This area appears to be flat in *Propithecus* but slightly dome shaped in *Indri*. Both these latter genera have the tapetum.

Tarsiidae are nocturnal animals. The retina of tarsiers has been studied by many students, not only because its visual system appears to be strikingly different from those of other prosimians, but also because of widespread interest in this peculiar primate. The retina of *Tarsius* is composed entirely of rods, as would be expected for a nocturnal animal, but it does have a *fovea* as well. However, it appears that this *fovea* is made up only of rods, thus, differing from the cone *fovea* usually found among primates. *Tarsius* does not have a tapetum. With these animals the density of photo receptors is much higher in the center of the retina than in the periphery.

Considered in an overall view, we note that the retinae of higher primates are characterized by an increase of the number of photo receptors toward the center of the retina. As already mentioned, all Anthropoidea show both types of receptors, rods, and cones except the night monkey, *Aotus trivirqatus*. Higher primates also always have a *macula lutea* and *fovea centralis* exclusively containing cones combined with areas of high visual acuity. This high acuity is possible because of a greater complexity of the synaptic relationships of the nerve cells in these areas. That is, in the center of the retina, more receptor cells are connected to more ganglion cells, and thus, the transmission of photo impulses becomes more individual and also more elaborate. Within the *macula* the cones are longer but thinner than in the outer retina. The retina is thicker and allows for higher resolution of the impulses received, and all these features result in higher visual acuity among Anthropoidea. It can generally be observed that in the retinal center, fewer receptors and connecting neurons are joined to one ganglion cell, and the absolute number of all cell units increases centrally. The differences between the retinae of various higher primates (except *Aotus*) are mainly quantitative rather than qualitative. The numerical relations between the different types of cells are different and often characteristic for genera or even species.

Equally important for the visual properties of primates such as visual acuity is the acquisition of stereoscopic vision. Visual impulses, caused by light that stimulates cones or rods of the retina, are transmitted by the optical nerve. Optical nerve fibers from the outside of the eyeball connect to subcortical integration centers of the brain hemisphere at the other side from the eye, whereas in animals with stereoscopic vision, some fibers of the inside of the eyeball also go to the cerebral hemisphere of the same side (Figure 7–11). This crossing-over of optical fibers takes place in the *chiasma opticum* (or optic chiasma), which is situated at the base of the midbrain just in front of the hypophysis. Among primitive mammals all the optic fibers are crossed (decussation) and go from either eyeball to the opposite sides of the brain. Because the eyes are situated laterally on the skull in such animals, two independent pictures are conveyed to the opposite sides of the brain. Thus, both images are evaluated independently. Perception of depth in space is

Figure 7–11 Optic chiasma.

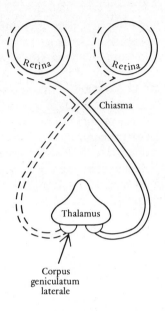

only possible in higher mammals, whose eyes are positioned more or less frontally and only a partial crossing of the optic fibers can be found. Each optical nerve, coming from the eye, carries fibers from both sides of the eye: the nasal half and the temporal half. The optical fibers from the temporal side of the retina remain uncrossed, and those of the nasal side cross over to the other optical tract. Thus, the optical fibers from the left half of both eyes and those of the right sides of each retina are carried together by the left and right optical tract respectively. The optical tracts terminate in the primary optical centers, the cerebral peduncles. In Anthropoidea and man about 40 percent of the fibers remain uncrossed and continue on to the optic tract of the same side as their origin and terminate in the same brain-side. The crossing of the optic fibers makes it possible for more or less identical optical impulses from both eyes to be evaluated in the same area of the brain. This fact makes stereoscopic vision possible. It is known that in *Tarsius* only 25–35 percent of the fibers remain uncrossed. The highly elaborate visual perception of humans seems to be closely correlated to our highly developed ability to evaluate optical impressions through the integration areas of the brain cortex (this ability is partly learned) rather than through differentiation of the primary optic receptors.

In primates most of the optic tract fibers terminate in the lateral geniculate body; it has a rather simple structure in tree shrews and is very elaborate in man. Neurons contained within the lateral geniculate bodies are arranged into layers. In primates two layers of large cells are always present, and a variable number of small cells that are also arranged in layers is found. All these layers are called laminae. In tree shrews the laminae are rather irregular in their arrangement and have a simple structure. Only one layer of small cells is found in addition to the two layers of large cells. Higher primates have a greater number of small celled laminae in addition to the two large celled ones. Three laminae have been reported for *Tarsius*; a total of six laminae was found in lemurs; and most Anthropoidea except *Aotus* and *Hylobates* show only four. *Saimiri*, like tree shrews, have only two layers. In lorises, *Microcebus* and tarsiers, the two laminae that consist of large cells are arranged in a convex pattern. The laminae are inverted and concave in most monkeys and slightly s-shaped in lemurs and man. Distinctive differences can be found in the way these laminae are arranged, in their cell density, and in the way the optic fibers terminate in them. All these structural reorganizations of the lateral geniculate body of primates are correlated to differences in neural activity and interaction that appears to be more elaborate in higher primates (Noback and Moskowitz, 1963).

In generalized mammals the axes of the eyes are directed laterally; in tree shrews the eye axes enclose an angle of 140 degrees. In most members of the family Lemuridae, this angle is reduced to between 60 and 70 degrees and measures only 30 degrees in monkeys. However, the axis through the center of the bony eyesocket does not usually coincide with the optical axis of the eye ball. It appears (Schultz, 1948) that the optical axes of the eye balls enclose a slightly smaller angle than the axes through the centers of the eyesockets. Among Anthropoidea the optical axes are usually directed more or less parallel to each other. The rotation of eyesockets and thus eyeballs into a more or less frontal position makes binocular (stereoscopic) vision possible, but on the other hand reduces the width of the visual field.

All primates have eye lids. In some of the Old World monkeys, the eyelids are colored brightly and contrasted to the coloring of the face. Thus, the eyelids in *Cercocebus aterrimus* are lightly colored, and those of *C. torquatus* are a striking chalky white. *Cercocebus cephus* is especially notable in this respect: the eyelids are colored in a vivid violet-blue and look very much like those of a fine lady with elaborate eye make-up. Among macaques the eyelids are strongly distinguished, being white and contrasted within the bare lightly brown face of *Macaca sylvana*, which also has a white skin area above the eyes. *M. fascicularis* has an area of white skin on the inner or nasal side of the lids. Contrasting, light-colored eyelids are also found in *Papio* and *Theropithecus gelada*. Eyelids with contrasting coloration function as signals when flashed by rapid movements of the lids. These "signals" are commonly intrepreted as threats. Very well documented is the threat-yawn of baboons with open mouth and closed eyelids, showing off an impressive set of teeth and the contrast-colored eyelids.

NUTRITION AND INTESTINAL TRACT

Primates are highly dependent on food sources that their environment offers. Interestingly, the intestinal tract seems not to be very much influenced by minor differences in the diet of mammals, but often shows characteristics that are group-specific. Highly arboreal primates have a diversified array of food items available to them like leaves, buds, blossoms, young shoots, bark, nuts, and fruit. Arboreal primates can also prey on insects and their larvae, on snails, and might occasionally eat an egg or even a young bird. Many primates have adapted to areas that are predominantly inhabited by man. There, primates can profit from man's cultivation of food products or even, at an extreme, from his garbage in the peripheries of cities (e.g. *Macaca mulatta*). *Macaca fascicularis* has become specialized in foraging in part on fruit, on insects, and on seafood. Thus, this primate is also commonly known as the "crab-eating macaque." *M. fascicularis* live predominantly in costal areas, catching crabs and other seafood in the shallow waters of the mangrove swamps lining the edge of the Malayan sea.

Some primates that live predominantly on the ground have become specialists in root digging. They eat green shoots of grasses and herbs as well as roots and bulbs, insects and their larvae, lizards, and also the seeds that are available from the grassland or bushes. Of all the prosimians, *Tarsius* lives most predominantly off animal protein: insects, larvae, and lizards. The indriid genera *Indri* and *Propithecus* are highly specialized herbivores and consequently, their bodies are rather pear-shaped, showing the enlarged gut of a bulk eater. Among monkeys, both the South American howler monkeys and the colobines of the Old World are highly specialized leaf eaters showing, however, different degrees of specialization in this dietary adaptation, as we shall soon see. The highland baboons—*Theropithecus gelada*—are specialized seed eaters (Jolly, 1970). Among apes the gorilla is herbivore, the orangutan predominantly a frugivore, and the chimpanzee an omnivore.

The migratory movements of most primates are closely related to the availability of food. Chimpanzees, for example, go every day to areas where certain fruit are ripe and abundant, returning every day until the supply is exhausted, and they have to find another area in which to forage. Both baboons and chimpanzees eat a variety of small mammals and antilope meat. Only chimpanzees actively hunt in groups (Jolly, 1972). In a recent report, Jane Goodall (1979) indicates active cannibalism and "warfare" among the Gombe Stream National Park chimpanzees. The cause of this hostility remains unexplained. Members of the subfamily Cercopithecinae have food pouches in their cheeks. These enlargements of the membrane of the oral cavity outside the teeth can be enormously-stretched. Such pouches are used as storage bags for food if it is abundant, or can be stuffed full of food in a hurry if the situation requires a hasty retreat, and then the food can be retrieved and eaten in peace at a hiding place. These pouches, when fully stuffed, can result in a different—even grotesque—appearance of an animal. Cheek pouches can extend far beyond the facial area, even bulging out and down into the neck. If the pouch is overstuffed, the animals often need to push the food items back out with the help of their hands because the cheek musculature is weakened by over-stretching.

We have already mentioned that the South American howler monkeys and also the Old World langurs (Colobinae) are highly specialized leafeaters. This specialization, however, is much more elaborate in the colobines than among howlers. Langurs have been found to have a ruminantlike digestion with sacculated stomachs (Kuhn, 1964; Bauchop and Martucci, 1968). This is not the case for howler monkeys that have comparatively large stomachs but have about the same ordinary basic anatomy as all other primates. The stomach of infant tree shrews is also comparatively very large. These animals nurse only at very long-separated intervals. Thus, the animals require a stomach that can store an unusually large amount of mother's milk. The majority of primates have stomachs with a simple structure, consisting of three anatomically and functionally different parts: the fundus, the body, and the pyloric portion. Also, the small intestine remains rather simple and uniform in all the primates. Many lower primates do not have a small intestine that can clearly be subdivided into duodenum, jejunum, and ileum. Rather, with these animals the small intestine appears to be nearly uniform throughout in shape and histology. Among such primates the entire small intestine is suspended dorso-medially by a mesentery and is not attached for the entire length between the pylorus of the stomach and the colon. This is the typical construction for tree shrews, tarsiers, lemurs, marmosets, and a number of cebids. With the remaining primates the lower part of the duodenum rides firmly attached by a special ligament to the posterior abdominal wall, the so-called *cavo-duodenal* ligament. In such cases that part of the intestinal area, where small and large intestine join, is proximal to the duodenum but below it. In lesser apes, great apes, and humans the duodenal portion of the small intestine is bound firmly to the posterior abdominal wall.

The large intestine or colon has mainly excretory functions; it shows certain interesting differences among primates. The large intestine appears less likely to be influenced by functional differences than is the case with the small intestine. With Tupaiiformes and Tarsiiformes the colon is very simple. In tupaias the colon continues more or less straight down to the anal opening, and the caecum—a blind-ended pocket at the upper end of the colon—is small. In tarsiers a very short transverse colon is found before it extends straight downward. *Microcebus* and *Cheirogaleus* have a colon that resembles that of tarsiers. Among Lemuridae and Lorisidae the colon is long and is turned into a number of corkscrewlike spirals or *ansa coli*. The colon is especially long and arranged in a spiral both in *Propithecus* (Figure 7–12) and *Indri*.

In most Old and New World monkeys the connection between ileum and colon is positioned deep—caudally—at the right side of the body, and as the colon ascends from there, it extends across the abdomen higher up and descends at the left side. A high variability of colon configurations has been found among individuals of the

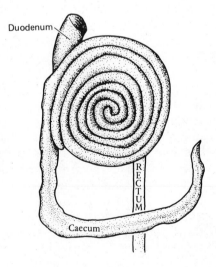

Figure 7–12 Disc-like colon of *Propithecus verreauxi* showing ventral disc. A second disc is positioned dorsally of this colon-disc and covered by it.

New World monkey *Saimiri*. Old and New World monkeys often have loops in the transverse branch of the large intestine. The ascending portion of the intestine is closely attached to the abdominal wall.

All the hominoid primates show a large intestine that is relatively shorter than in monkeys. Moreover, there are no loops in its transverse portion. Both ascending and descending branches are closely attached dorsally to the abdominal wall. The intestine terminates in the short rectum that leads into the anal opening.

A blind-ended extension of the upper colon sticks out from the junction of small intestine and colon. This blind sack has the same diameter as the colon and is called the caecum; it is found in all primates. The sack can often be hook-shaped or even slightly spiral shaped. An additional blind extension from the end of the caecum itself with a much smaller diameter is called the *appendix vermiformis*. This "true appendix" contains lymphatic tissue, the function of which is not quite understood. The *appendix vermiformis* is found in all Hominoidae.

The liver and spleen also show various characteristic differences in primates. In prosimians the liver is usually differentiated into multiple lobes, much more divided than among higher primates. Liver and spleen do not have a very well-defined shape of their own. Their shapes change easily, depending, among other things, on size and form of the adjoining organs such as the stomach. The spleen shows specific configurations in different primate groups. Each of the following groups shares common traits of the spleen: Tupaiidae, Prosimii, Cebidae, Callitrichidae, Cercopithecidae, Colobidae, and Pongidae (Starck, 1960).

chapter 8

Placentation, Reproduction, and Growth

Initially, it should be stressed that the evaluation of unitary sets of characteristics like placentation with the Order Primates (just as in any other group of animals) and their conversion into phylogenetic and/or systematic reconstructions are of little value by themselves. Such character complexes must be brought together with findings from all the other sources of taxonomic data known for a particular group of animals.

An especially intricate variety of structures, substructures, and modes of function are those that are effective during ontogeny. This fact is particularly true for the process of placentation, and it requires considerable knowledge to allow for the understanding and evaluation in a taxonomic sense. Many parallel developments and functionally based adaptations can be found during the very early growth stages of placental mammals.

Within some orders of mammals, placental relationship can be recognized in early ontogeny. Thus, for example, in carnivores the mode of placentation is very uniform; this is also true for Orders Cetacea, Artiodactyla, and Perissodactyla. Within the Order Primates, however, we find a surprising range of differences and differentiations during early ontogeny. Thus, according to Starck (1956), it seems that differences in mode of placentation between the two suborders of primates do at least crudely reflect the evolutionary "grades" of these suborders. Some groups of primates can be distinguished by their early ontogeny and placentation:

Tupaiiformes, Lemuridae together with Lorisidae, Tarsiidae, Cebidae, Cercopithecoidea, and Hominoidea. These findings, however, do not give much information about the relationship of these major groups of primates to each other. Even today, data are rather scanty on which the grouping of the primates according to their mode of placentation could be based. Some data are still contradictory. Decisions should perhaps be left to those scientists who can base their knowledge and their opinion on ample experience in this field when drawing conclusions from primate placentation. Furthermore, it seems mandatory to rely on those scholars whose knowledge is not restricted to research on placentation and early ontogeny of primates alone, but who also have knowledge about these stages of development throughout the mammalia. And there is, for instance, a large literature on the subject of primate placentation in German and French that is not often cited in English.

EARLY PRIMATE DEVELOPMENT

Should a fertilized egg wander into the uterus, it finds the inner uterus lining or endometrium (also called decidua) ready to receive the early blastocyst. The lining has become thick and highly vascular, its connective tissue loosened, all the capillaries plump, and the glandulae active. The outer layer of the germ or blastocyst (also called trophoblast) "digests" the endometrium and burrows into the uterine wall. At the same time the germ begins to build up the functional relationship with the maternal organ that also reacts to its presence. If no fertilization took place, the inner lining of the uterus is shed after about twelve to fourteen days following ovulation in humans. This process is also true for most primates, and only the duration of the cycle varies somewhat. (Primate female cycles usually vary between twenty and thirty-five days in length—Jolly, 1972). Then the preparatory cycle of the female reproductive system starts all over again. If an egg is fertilized and moves into the uterus where it implants itself into the wall, the fetal membranes and placenta begin to develop their intricate correlation. The blood circulations of the two individuals always remain separated from each other. However, in the area of contact between the two, exchange of nourishment, oxygen, and fetal waste takes place. Also, the placental membrane can pass certain amounts of amino acids and albumins back and forth. Other substances, for instance, certain antibodies or viruses, however, cannot pass through.

We know that among insectivores the mode of placentation and early ontogeny is by no means uniform. It is, for example, possible in certain insectivores that two different kinds of placental structures of different functions can occur in one individual. Such knowledge also emphasizes the importance of taking very critically taxonomic judgments based on primate placentation.

In general, it should be said that the placenta is a highly complicated organ that regulates the metabolic exchange of nourishment and excretion; it regulates breathing and all the fundamental biological functions and vital interrelationships between the mother and the unborn offspring. The degree of contact between the two individuals can be very different in different types of placentation. These differences are the basis for the attempts to interpret taxonomic and phylogenetic relationships of primates that will be discussed in what follows.

The early embryo is situated within an enveloping membrane called the serosa or also the chorion (*chorion* is Greek for "skin"). At a later stage a membranous fold grows out and around the embryo. The edges of this fold, called the amnion fold (*amnion* in Greek is "sheepskin") approach each other and finally fuse together. The result is that the embryo is surrounded by two layers of cellular material, namely the amnion immediately surrounding it and the serosa or chorion outside. The chorion is the organ that maintains the connection between the maternal uterus and the offspring. Inside the chorion we find a baglike structure that originates from the embryonic rectum. This protrusion of the embryonic rectum functions as a bladder that is commonly called the allantois. This allentois always remains connected with the embryo during its dependent life. It is the allantois through which the vessels pass from the embryo in order to enter the area of the chorion where the offsprings structures meet the maternal structures. The chorion is also called the trophoblast. It forms numerous protrusions that enlarge the contact area with the maternal uterus. The embryonic vessles intrude into those chorion protrusions that in turn connect closely with the maternal uterus. The placenta grows into different shapes and sizes around the embryo in different primates. Among primates we can distinguish morphologically between two basic chorion types (and thus two placental types):

1. Chorion protrusions cover the entire placenta: *placenta diffusa,* usually found in epithelio-chorial placentae
2. Chorion protrusions occur only in a disclike area of the placental surface: *placenta discoidalis* seen in endothelio- and hemochorial placentae.

In the latter case one should distinguish between two subtypes: A. Simple *placenta discoidalis* with one disc of protrusions only, and B. Placentae with two discs of protrusions that are called *placentae bidiscoidales* (*bi* in Greek means "two"). There are several other types of placentae found in various other mammalian orders, but they will not be mentioned here because they do not occur in primates.

Apart from these macroscopic morphological differences of placentae, one can also distinguish between the different degrees of intimacy between embryo and uterus that are caused by differences of the substructure of the two individuals concerned. Thus, it is possible that the connection between the embryonic protrusions of the placenta and those of the enveloping uterine material are not closely united but only touch each other. Both will separate from each other easily at the time of birth. This condition of the placenta is called an adeciduate or nondeciduate placenta. The innermost lining of the uterus is called membrana decidua or, more simply, just the decidua. In the case just described, when at birth the uterine lining remains unruptured, the placenta is called adeciduate because of this condition. It is, however, also possible that the two individual portions of the placenta can be intricately connected with each other and that after expulsion of the offspring and during removal of the placenta, the inner lining or decidua of the uterus does not easily separate from the placental portion and is shed with it. In these cases the uterus suffers a rupture, and because the inner lining of the uterus that is expelled is called the decidua, we speak of this condition as a placenta decidua. Nondeciduate placentae are usually of the placenta diffusa type in primates.

During early ontogeny the tissues that originally (and always) separate the fetal vessels from the maternal uterus vessels can undergo resorption. Thus, it happens that the number of tissue layers between the two individuals is reduced, and consequently the interchange between maternal and fetal bloodstream becomes more efficient and intimate. Not only does reduction of the number of layers occur, but also single layers of tissue may be reduced in thickness. Accordingly, the intensity of physiological exchange (interchange) between mother and offspring differs in different types of placentation. The following typology of placental types is based on an attempt to categorize morphological differences of placentae, as proposed by Grosser (1909); this typology deals here only with those types that are found in primates.

In cases where all the tissue layers of the uterus and all those of the chorion are present, we find the following series of six layers that separate the maternal blood flow from that of her offspring:

Mother:	1. Wall of maternal vessel
	2. Layer of connective tissue
	3. Uterus epithelium (decidua)
Offspring:	4. Chorion epithelium
	5. Connective tissue of chorion
	6. Wall of fetal vessel

If this series is complete, we call the placenta epithelio-chorial. This is the case in any early stage of eutherian ontogeny. If layers 2 and 3 are resorbed, and the epithelium of the chorion comes to connect directly with the walls of the maternal blood vessels, the placenta is called an endothelio-chorial placenta. Among primates we also find that the maternal layers 1 through 3 undergo resorption. Then the surface of the chorion is directly exposed to the maternal blood stream. This latter type of placentation is called haemo-chorial and allows the most efficient interchange between mother and fetus.

All these three placental types can be said to be present (in a descriptive sense only) within the Order Primates. Always, it has to be kept in mind that similarities in placentation can be caused by parallel developments of functional similarities and do not necessarily reflect either taxonomic or phylogenetic relationships. Also, a similarity of full-term placentae can be reached in very different ways; for example, according to Starck (1956) the placentae of *Tarsius* and higher primates look very similar in their final stages but develop in very different ways. Differences in placentation can also be caused by differences in body size.

As already mentioned, five higher categories of primates the Lemuridae together with Lorisidae, the Tarsiidae, the New World monkeys, the Old World monkeys, and finally the Hominoidea appear to be groups of rather uniform placentation (within each group). We would go too far in this context to discuss all the intricate details of the early development of fertilized primate eggs and their mode of implantation and fetal life, but see Starck (1956).

Tupaiidae

Within all the genera of this family where it is known, the mode of placentation is the same in spite of the fact that there are considerable differences in body size between genera (e.g. *Tupaia glis* averages 170 gr, *T. minor* only 50 gr). It has formerly been believed—as published by van Herwerden (1906)—that the placenta of tree shrews is not expelled but undergoes resorption *in situ* after birth. This situation is called contra-deciduate. That the situation occurs in tree shrews has been widely quoted in the primate literature, but—as Starck (1974) points out—seems not to be the rule, if true at all.

The structure of the tree shrew placenta has been described by Luckett (1968) as being endothelio-chorial. The tree shrew placenta is bidiscoidal; but it is different from the true primate bidiscoidal placenta because that of *Tupaia* has two separate branches of the umbilical cord that lead to two discs of equal size. These umbilical branches split off from the main trunk of the umbilical cord and insert centrally. In bidiscoidal placentae of Anthropoidea, the umbilical cord goes directly to the first and larger placental disc, and an anastomosis connects from there to the second and smaller placental disc, a condition quite unlike the situation in tree shrews.

Lemuridae and Lorisidae

The placenta is nondeciduate, diffuse, and epithelio-chorial. However, *Galago demidovii*—the dwarf galago—was investigated by the French scholar Gérard (1932) who describes its placenta as having a well-defined region of endothelio-chorial type within an epithelio-chorial placenta. This discovery has been widely accepted by subsequent authors but have never been reconfirmed by someone else, presumably because of lack of material. Starck (1956), however, thinks that Gérard's findings might prove to be correct, and that therefore it is of interest to keep in mind that the dwarf galago may have an area of higher functional interchange within a placenta of otherwise lemuroid character.

Tarsiidae

The early ontogenetic events in *Tarsius* are the same as in all the other prosimians. Implantation is not different from that of lemuroids and lorisoids. During later stages of the intrauterine development, unique specializations of the fine structure are characteristic for tarsiers. Only the full-term placenta resembles that of higher primates in being haemochorial, discoidal, and deciduate. Therefore, it seems not justifiable to state that the placentation of tarsiers and monkeys would be identical (see Starck, 1956, 1974, contra Luckett, 1968). The ontogeny of the tarsier placenta is very specialized, differs from that in higher primates, and connot be grouped together with any other haemochorial mammalian placenta.

Ceboidea and Cercopithecoidea

In most anthropoid monkeys the chorion establishes two contact zones with the uterine wall and consequently, the placenta is bidiscoidal. This is the case in all those New World primates whose placentation is known, except *Alouatta*—the howler monkey. Among Cercopithecidae the placentae of the genera *Macaca* and *Cercopithecus* are usually bidiscoidal, which is also the case in those Colobidae that have been studied. All the specimens of the genus *Papio* that have been analyzed have discoidal placentae. All these Old World monkey placentae are also haemo-chorial and deciduate. The two discs of the bidiscoidal placental type are different in size and are also formed at different times, one after the other. The first one differentiates when close to the embryonic pole and soon becomes larger than the second trophoblast that later fuses with the opposite uterine wall. Structurally, the two placental discs are alike. Anastomoses of vessels within the chorion connect the two placental areas. In New World primates the placenta forms comparatively later in ontogeny than in Old World monkeys, and the implantation is slower. The Ceboid placenta also never becomes as elaborate and efficient in its functional interrelationship with the embryo than the placenta of Cercopithecoidea. The mature placentae of Cercopithecoidea and Ceboidea are very similar to each other, though their ontogeny before they reach this final stage is different, especially in terms of timing. The umbilical cord of higher primates, as already mentioned, is different from that of the bidiscoidal tree shrew placenta because it leads to the first and larger placental disc without branching. The two discs are connected by anastomoses from the first placental disc.

The placentation of New World monkeys can be regarded as more primitive and simple that that of Old World monkeys and Hominoidea, but it can be considered as a possible model for an earlier phylogenetic stage leading to the latter. However, this does not mean that the Ceboidea are ancestral to the Old World primates.

Pongidae and Hominidae

The placenta is discoidal, haemo-chorial, and deciduate. Structurally the mature placentae of Pongidae and Hominidae resemble those of the Cercopithecoidea. They are, however, very different from those in their early stages of development. Implantation of the fertilized egg is deep within the mucosa of the uterus. Consequently, the subsequent development of the placentation is different from that in Cercopithecoidea and Ceboidea, in which the placentation starts with a rather superficial implantation. Starck (1974) points out that the difference in depth of implantation causes the basic differences between the placentation in New and Old World monkeys on the one side and in man and apes on the other. Thus, the differences arise gradually and are not structural.

It can be assumed that the Hominoidea passed through an earlier placentation stage that was very much like that of present-day Cercopithecoidea, and that both Hominoidea and Cercopithecoidea might have previously passed through a "ceboid stage" of placentation.

Major theories about the phylogeny of primate placentation have been presented. Very thorough discussions of the material have been made in German by Starck, 1955, 1959, and 1974. Martin's hypothesis (1968) on the possible evolution of placentation among primates is essentially the same as that of Starck formulated about ten years earlier.

REPRODUCTIVE ORGANS, REPRODUCTION, GROWTH, AND DEVELOPMENT

Adult male primates are characterized by permanently descended testicles. It appears that in the tree shrews under stress, the testicles (usually extra abdominal) can be temporarily retracted into the abdomen. All male primates have a pendulous penis. The scrotal sac that envelops the testicles is located at the sides of the penis as in Tupaiidae, Callitrichidae, the gorilla, and the Hylobatidae. In the latter family the sac is sometimes even positioned on top or over the penis. The scrotum is positioned underneath the penis and more or less independent of it in all other primates. The penis of many primates contains a so-called baculum, or penis bone, as is also the case in many other mammals. This baculum is missing among Tupaiidae, Tarsiidae, the cebids *Lagothrix* and *Ateles*, and in man. No correlation has been found between the size of the adult males and that of their bacula. Schultz (1969) measured the baculum of a full grown *Gorilla* that was only 11 mm in length, whereas those of a mandrill and potto measured 23mm and 21 mm respectively. As a rule the baculum of prosimian primates is relatively large. Also, testicular size has no correlation to body size in adult male primates (Schultz 1938). Large testes are the rule in primates, whose females have large sexual swellings as, for example, chimpanzees. Adult male chimpanzees can have testicles that together weight up to 250 grams, and those of a wild shot healthy male gorilla weigh no more than 36 grams. Seasonal changes in testicular size are known in *Varecia, Microcebus,* and *Cheirogaleus.* Also, some macaques have been shown to have seasonal changes in testicle size.

Some female primates (both prosimians and New World monkeys) have a bony structure in their clitoris that is smaller than the male baculum. Many prosimians have spurlike projections on the glans penis. Among higher primates such differentiations are only known in *Ateles.* Some female prosimians and cebid monkeys have an enormously long clitoris resembling the penis that makes sexual evaluation at a distance difficult. The testicles descend in most primates not long after birth, but the descent is usually earlier in higher primates than in prosimians. The scrotal skin contains special kinds of glands in *Perodicticus, Cebuella, Oedipomidas,* and many Malagasy lemurs. These glands are distinctively colored in *Cebuella.* In some marmosets the testes can descend deeper into the scrotal sac as these animals make threat displays, and the intensely colored scrotal area becomes even more obvious (Epple, 1967).

In *Cercopithecus* species, in *Papio sphinx,* and in *P. leucophaeus* the scrotal skin is brightly colored. The skin is blue in many of the guenons, especially so in the *Cercopithecus aethiops* group and is bright red in Allen's guenon—called Allen's swamp monkey (*C. nigriviridis*). Among mandrills the circum-anal area is red. Lateral to their ischial callosities, the skin is bright blue grading into bright purplish-red at the outside. In the subgenus *Colobus (Pilocolobus),* swellings of the area between ischial callosities have an intensively shining pink color in juvenile males. This coloration, however, fades away with age. These pink structures in young male *Colobus badius* have been interpreted as a simulation of the female perianal area and are said to play a social role (Kuhn, 1967). The scrotal area of mandrills exhibits a rainbow of iridescent colors: the back of the scrotum is purplish-red, the front brightly pink, the pubic region is scarlet, and the glans penis intensely pink. In the drill an analogous coloration of this area appears to be even more intense in the color hues, and there is a metallic shine to the colors.

As already mentioned, homologous to the penis in males is the clitoris in females, which is usually small and hidden between the *labia majora.* In some prosimians and some of the New World cebid species the clitoris can be very large and pendulous and even larger than the penis among males of the same species. Among Tupaiiformes and Lorisiformes the urethra opens at the very tip of the clitoris. In all other primates the urethra opens near the base of the clitoris. *Labia majora* are found in Cebidae, Cercopithecidae, the lesser and greater apes and in man.

The uterus develops ontogenetically from paired tubes (Müller's tubes). These tubes fuse at the lower end to a larger hollow organ, the body of the uterus, and the extent of the fusion varies in different primates. If parts of the two tubes remain separate, they are called "horns," and a uterus that retains separate horns of some length is called a *uterus bicornis* or double-horned uterus. In tree shrews the body of the uterus is relatively short and the two horns remain separate and long. Also, in *Tarsius* and the members of the Lorisidae the two horns are rather long. To a lesser extent separation of the horns is also maintained in all other prosimians, but they are very short in most other prosimian primates.

Among all Anthropoidea the tubes fuse totally and only one simple uterine body is present, which is called the *uterus simplex.* Only in some of the marmosets, a slight separation of the upper end of the uterus by a medial indentation can be observed.

Sir Solly Zuckerman gained considerable attention when, in 1932, he stated that sex plays a crucial role in primate societies. He emphasized that male primates were always ready to mate with females that equally constantly were sexually responsive to the males. Zuckerman came to this conclusion after studying the social and sexual behavior of captive monkeys and chimpanzees and also from his knowledge of human behavior. It is true that many anthropoid primates are sexually receptive all the year round when in captivity. Also, humans do not seem to have any obvious seasonal restrictions or mating seasons. However, most primates have breeding seasons in the wild. Many species also continue to have breeding seasons when in captivity.

Mating seasons and consequently, birth and breeding seasons are restricted to only a couple of days or even hours in many prosimians. Even in captivity associated lemur females come into estrus synchronously within hours of each other and have an equally short birth season. This condition can be the case in animals of the same species, even when these animals have neither visual nor olfactory contact with each other. However, the physical location of these animals does influence the timing of these very limited reproductive periods of lemurs. Highly synchronized mating and breeding occur particularly among ring-tailed lemurs (*Lemur catta*) and in the New World monkey *Saimiri*. Both these species engage in heavy scent marking during the mating season, and it may be that sexual pheromones play a decisive role in the coordination of the receptivity in females that come into estrus at the same time (compare Jolly, 1972). It appears that in many lemurs, the highly restricted birth season is timed so that the infants are weaned from their mothers during the wet season in Madagascar, thus, assuring the infants an abundance of suitable food at a crucial stage of development (Petter-Rousseaux, 1968). Also, tupaias have a definite breeding season in the wild. Many of the seasonal breeders that continue to breed only during restricted time periods in captivity do so with high dependency on their geographical location. Thus, *Microcebus* gives birth during November through February in Madagascar and during May through July when kept in the Northern Hemisphere, where seasons are reversed. Wild tarsiers seem to breed during the entire year in the wild: Newborn animals have been reported in all seasons. Even among higher primates that seem not to have well-defined breeding seasons, births seem to occur predominantly at a certain time of the year. Marmosets are known to have breeding seasons in the wild. Cebidae, however, evidently do not have restricted birth seasons.

Presbytis apparently breed throughout the year in the wild. The same is true for many *Cercopithecus* species. Other cercopithecines have been reported to show a concentration of births during seasons with ample food supply. *Macaca mulatta* newborns are most abundant during March and April in India, and only a few occasional ones are born as late as September. Mating occurs predominantly during January through March among Japanese macaques, and the birth season there runs from early June until the middle of August. Generally, baboons seem to have seasonal peaks of mating and breeding but no clearly defined seasonality. These peaks are much more precisely restricted in *Papio hamadryas* of Ethiopia. This condition is also true for *Cercopithecus aethiops*, which are within the same general geographic area. There in Ethiopia, the maximum incidence of birth is in April and May. Most species of *Colobus* as well as the lesser apes and great apes are thought not to have such seasonality. The data gathered so far concerning mating and birth seasons in the wild are still comparatively scanty and thus inconclusive concerning these animals. Female primates go through estrus cycles of twenty to thirty-five days duration. However, ovulation of fertile eggs does not necessarily occur continuously.

In many prosimians the vagina opens only during the few hours or days of estrus—the time of highest receptivity of the female.

Slight and very irregular menstrual bleeding has been observed in captive tarsiers and is reported for certain in New World monkeys. Periodic menstrual bleeding is of common occurrence in Old World monkeys. A feature that is restricted to the Old World primates is the marked change in the circum-anal and genital areas that occurs during ovulation. This period is also the time of highest sexual receptivity, when a ripe egg is ready for impregnation. Sexual swellings are often accompanied by color changes in baboons and among mangabeys and are very pronounced in the chimpanzee. Swellings also occur in *Cercopithecus talapoin* and several species of *Macaca;* however, these swellings are less pronounced in their appearance than those of baboons, mangabeys, and chimpanzees. It has been reported that *Colobus (Pilocolobus) badius* females have sexual swellings similar to those in baboons, extending into the base of the tail, and that they become permanent enlargements of adult females (Kuhn, 1967). This latter happening, however, invalidates the sexual swellings and colorchanges as signals of the fact that the females are in estrus in this species. The macaques include some species that show only slight changes during estrus, such as *Macaca mulatta, M. maura,* and *M. assamensis.* The females of these species turn red in the genital area, at the root of the tail, and upon thighs and chest. Even the face turns red. However, these color changes are accompanied only by slight swellings of the genital regions. In *Macaca sylvana* the circum-anal swellings and also the rump turn bluish gray rather than red, and the swellings are slate gray in *M. fascicularis.* Many female macaques also have a strong odor during estrus caused by mucous vaginal secretion. In orangutans swelling of the sexual skin can occur during pregnancy but not during ovulation. Chimpanzees have very large sexual swellings that are bright red during estrus and that can reappear during pregnancy. The protruding sexual turgescence of female baboons sometimes ceases to subside between ovulations in old animals. This condition can, as Schultz (1969) puts it, "attain such grotesque proportions that the poor creature can no longer sit down."

Most primates have to learn how to mount and copulate successfully. This fact has been emphasized by the demonstrated inability to copulate of animals that have been brought up under conditions of isolation (Harlow and Harlow, 1962). Among most lower primates the male mounts the female from the rear and embraces her around the chest. Sometimes, male monkeys grasp the ankles of the females with their feet and thus rest their entire weight on the female during copulation. Pottos have been observed copulating ventrally, facing each other. Rear copulations play an important part in contexts other than reproduction among social primates. Only among Hominoidea do copulatory positions become quite variable.

The number of offspring at a given birth varies somewhat among tupaias and the primates. Tree shrews and the dwarf and mouse lemurs have up to four newborn in one litter. However, the average litter-size for both these species is two. Among marmosets and ruffed lemurs twin births are the rule, but triplets are also relatively common. Usually, two ripe eggs are released at ovulation in marmosets and thus, their twins are fraternal. The multiple births of prosimians also seem to be fraternal in the vast majority of cases. All other primates have predominantly single births, but twinning occurs in every five to six births in *Lemur catta* and some races of the brown lemur (*L. macaco fulvus*). Among higher primates twins are born in about the same percentage as in man, namely in approximately one out of a hundred births and, as in human twins, 80 percent are fraternal.

Glaser (1970) was able to show that most prosimian primates and many New World monkeys tend to have the same level of maturation in their skeletal ossification at birth, as does *Homo sapiens*. It turns out that *Hylobates* and *Gorilla* among the apes and *Nasalis* (one genus of the Colobinae) are somewhat ahead of the prosimians, New World monkeys, and man in terms of bone-maturation at the time of birth. Thus, *Hylobates*, *Gorilla*, and *Nasalis* have an intermediate position in this respect, whereas all the representatives of the subfamily Cercopithecinae that have been studied show the highest degree of ossification at birth among primates. It appears reasonable to say that all primates are dependent on their parents to a certain degree immediately after birth and to a lesser extent for an extended amount of time thereafter. Even though the Cercopithecinae show a higher degree of physical maturation at birth than the other primates, their dependency on parental care lasts at an average six months.

The following stages of life can be recognized:

fetal = intrauterine phase
infantile = from birth until the eruption of the first permanent molar
juvenile = from eruption of first permanent molar until permanent dentition is completed
adult = from completion of the permanent dentition on

Growth does not only concern the general increase in body size but also those slowdowns of growth rates in some regions of the body that result in proportional differences in various parts of the body. Length and intensity of growth vary widely among primates. Rates are closely related to the absolute size of the adult animals and thus, also sex related, resulting, for example, in differences in body size and proportions of males and females. Different parts of the body grow at different rates. Thus, during prenatal life, growth occurs with a pronounced cranio-caudal gradient, resulting not only in larger size of the cranial parts of the body but also in a more advanced degree in their developmental detail at birth. Fetal primates are more similar to each other than are adults. The absolute length of the different stages of life generally increases among primates from prosimians through monkeys, lesser apes, great apes, and man. The adult stage is especially long in modern man but, as such, can certainly be interpreted as an artifact—a product of improved general and medical care. Also, nonhuman primates in captivity appear to live longer, having profited from the same medical improvements as man, and as the understanding of their housing and dietary requirements increases (Jones, 1968).

The fetal period lasts only six weeks in tree shrews, nine weeks in *Microcebus*, nine or ten weeks in *Cheirogaleus*, around eighteen to twenty-four weeks among lemurs, and twenty weeks in most bush babies. In New World monkeys gestation length usually ranges around twenty to twenty-five weeks. Among the Old World monkeys macaques are born at twenty-four and baboons at twenty-seven weeks. Lesser apes are reported to need thirty weeks until the fetus is ready to be born. The chimpanzee has an average gestation length of thirty-four weeks, and thirty-eight weeks seem to be the average for orangutans and gorillas. Also, human beings have a gestation period of about thirty-eight weeks. From these data it becomes obvious that gestation length is not correlated with body size.

According to Schultz (1969), the timing of the birth is most closely defined by the size of the full-term baby and the dimensions of the maternal birth canal. Leutenegger (1973) also analyzed the factors governing size correlation between the diameter of the maternal birth canal and the headsize of full-term primate fetuses. His very interesting approach must be somewhat tempered when evaluation the role that locomotor requirements have in shaping the maternal birth canal because he bases his observations on locomotor groups that are not valid. For example, Leutenegger lumps howler monkeys together with woolly and spider monkeys as "New World semibrachiators." This observation is, at the least, an oversimplification, if not quite wrong (Grand, 1968, Mittermeier and Fleagle, 1976). Also, some of his other locomotor classifications weaken his interpretation. Nevertheless, the basic thoughts underlying Leutenegger's reasoning are worth considering. A revised study will, we hope, yield a clearer picture of his basic assumptions concerning the effect of locomotor influences on the pelvis. Certainly, general body size, head size of the full-term fetus, and locomotor forces shaping the birth-canal of the maternal pelvis do have a part in determining the time when the fetus will be born. However, it seems that these are not the only factors that are of importance in this respect and that the events leading up to parturition are more complex. Our data still are rather scanty. Only in macaques, chimpanzees, and man do we have detailed studies of prenatal growth and development. It was Schultz who suggested that there is a successive increase in duration of all the main lifestages running from lemurs, through macaques, gibbons, and chimpanzees to man. He showed this in a diagram that subsequently has been widely republished and elaborated by others. From this diagram one would conclude that in fact the different stages of life increase gradually in the array of primates mentioned above. More recent data have changed the picture. It turns out that lemurs can live close to thirty years without showing any extreme signs of aging and can also continue to reproduce. Schultz thought that chimpanzees had a life expectancy of about thirty-five years. Today, we know that the life expectancy for a chimpanzee is closer to forty years, even in the wild. Individual chimpanzees have lived to be fifty and even sixty years of age (Riopelle, 1963). Chimpanzees can still reproduce at the age of forty years. Already, among prosimians, compared with other mammals, the different stages of life have increased in length. As Dukelow, (1978) shows, the longest gestation period in prosimians (187 days) so far reported for *Nycticebus*—the slow loris—is longer than that of the New World monkey *Cebus* (180 days), the Old World monkeys *Papio* (183 days), *Cercocebus, Macaca, Erythrocebus, Nasalis,* and *Cercopithecus* (147 to 167 days average). This observation certainly suggests that behavioral and social factors also have influence on gestation length rather than an animal's position in an organizational "ladder" based on absolute body size or taxonomic proximity to man. Long gestation periods do space out offspring and tend to decrease the total number of offspring. Hofer (in Hofer and Altner, 1972) points out that prehistoric man and today's chimpanzee may well have had quite similar reproductive patterns in terms of the spacing of their offspring, maximum age, survival of adults (grandparents), and length of the different life stages. However, if we look at total life length and the different periods—gestation length, infancy, juvenile stage, and adulthood—as percentages of the entire life span, it turns out

Placentation, Reproduction, and Growth **243**

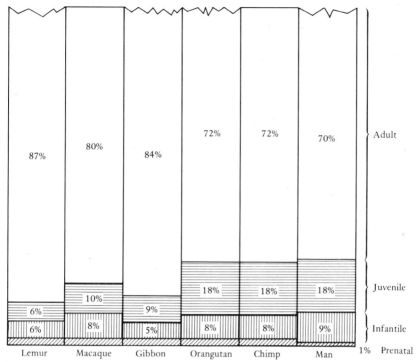

Figure 8–1 Diagram of lifespans with the entire length of life regarded as 100% and the different stages expressed as percentages of the lifespan.

(Figure 8–1) that these are very uniform among primates. Only the juvenile stage is somewhat longer relative to the entire length of life in monkeys and lesser apes if compared to the length of this stage in lemurs, and this stage is markedly longer in chimpanzee and man; we presume that learning has become a crucial factor in the adolescent development of primates.

Nevertheless, we do not have comprehensive data about gestation length, life-span, and the time duration of the different life stages in very many primates. Consequently, generalization must be a crude simplification of the actual facts at this stage, and the true story of primate developmental history remains beyond our present knowledge.

chapter 9
Locomotion

One of the basic activities of vertebrates is their locomotion: their manner of getting about. Because part of this activity is postural, the topic includes resting behavior. How animals move about is influenced by numerous factors. These are both internal, such as hunger, and external, originating from the surrounding world. External modifiers of locomotion can include many things, for example, social interaction with conspecific partners, actions caused by predators, or alterations of the environment.

Also, body size, limb length, limb construction, acuity of sensory perception, type of environment, and climate all play important parts in relation to locomotion. Evolution has produced a high correlation between environmental substrate, morphology, and locomotor adaptation. This is well illustrated by the common fishlike shape of water-living mammals, such as whales or porpoises, or alternatively, by the birdlike bodies and wings of bats. But these mammals live in extreme environments, and therefore, they are extreme morphological cases. Primates, compared with whales and bats, do not live in extreme environments. Thus, primates remain relatively unspecialized in terms of the morphology of body and limb. They have retained from early mammals many of the features of the locomotor apparatus often lost in other mammal groups, for example, pentadactyl hands and feet as well as clavicles. One of the most successful mammals, *Homo sapiens*, is unique in terms of habitual locomotor type compared to all other mammals. One could even ask: Is human success as an organism in any possible way correlated to our unique way of moving about? Answering this question could contribute a basic element to our self-understanding.

Within the Order Primates we know of a wide variety of locomotor habits. Most nonhuman primates spend at least some time during the day on trees. The majority rarely come down to the ground at all. Grasping when climbing in trees is an essential concomitant of life there and having pentadactyl hands and feet facilitates holding on. Moreover, grasping abilities allow a wide range of locomotor activities.

The works of many of the early researchers show that locomotion has long been considered important in the study of primate evolution (Keith, 1902; also Wood-Jones, 1929; and Gregory, 1934). Nevertheless, such discussions were for the most part theoretical because until recently very little was known about the locomotion of extant primates in their natural habitats. Information about primate locomotion came from observations on captive animals or from sketchy reports on accidental observations in the field. These tales were often exaggerated and certainly not always accurate. It is understandable then, that the overall anatomical similarities of the Hominoidea in the morphology of trunk and forelimb led to the long-lasting and widespread belief that the great and lesser apes all practiced the same locomotion—"brachiation"—and that we have originated from arm-swinging ancestors. "Brachiation" is a type of arboreal locomotion where progress is made by propelling the body forward with the arms, which are extended above the head. When arm swinging locomotion is rapid, both hands are free between alternate grips. We now know that only the lesser apes habitually brachiate to any great extent. As will soon be discussed, the great apes have their own distinctive modes of locomotion. This demonstrates that an overall similar morphology does not necessarily mean similar locomotion. Before analyzing the locomotion of a fossil animal on the basis of the morphology of its remains, one needs to know the relationship between the anatomy and the locomotion of extant forms that are being used as analogies.

The vast majority of field research on primates has been done since 1960. Initially, primatologists collected information primarily about social behavior and tended to neglect study of locomotion of primates. Since 1967, however, a number of comprehensive field studies have been made on the locomotion of primates in natural habitats. Thus, a better basis for our understanding of the locomotor behavior of different primates is slowly unfolding. Even though we now have quite a number of careful observations of the locomotor activities that primates exhibit in the wild, our records are still far from complete. It also becomes more evident than ever before that the mode of locomotion of a group of primates is highly dependent on the structure of the environment that they inhabit. Many references to papers concerning primate locomotion can be found in Morbeck et al. (1979) and McArdle (1981). All these studies provide information essential to understanding the relationships between morphology, locomotion, and habitat utilization. In 1964 Ashton and Oxnard introduced a classification of primates based on locomotion. In 1967 Napier and Napier elaborated and perfected this classification into locomotor groups (see Table 9–1).

Table 9–1 Locomotor Classifications Proposed by Napier and Napier (1967)

Category	Sub-Type	Activity	Primate Genera
1. Vertical Clinging and Leaping		Leaping in trees and hopping on the ground	Avahi, Galago, Hapalemur, Lepilemur, Propithecus, Indri, Tarsius
2. Quadrupedalism	(i) Slow climbing type.	Cautious climbing—no leaping or branch running.	Arctocebus, Loris, Nycticebus, Perodicticus
	(ii) Branch running and walking type.	Climbing, springing, branch running and jumping.	Aotus, Cacajao, Callicebus, Callimico, Callithrix, Cebuella, Cebus, Cercopithecus, Cheirogaleus, Chiropotes, Lemur, Leontideus, Phaner, Pithecia, Saguinus, Saimiri, Tupaia

Table 9-1 (continued)

Category	Sub-Type	Activity	Primate Genera
	(iii) Ground running and walking type.	Climbing, ground running.	Macaca, Mandrillus, Papio, Theropithecus, Erythrocebus
	(iv) New World semi-brachiation type.	Arm-swinging with use of prehensile tail; little leaping.	Alouatta, Ateles, Brachyteles, Lagothrix
	(v) Old World semi-brachiation type.	Arm-swinging and leaping	Colobus, Nasalis, Presbytis, Pygathrix, Rhinopithecus, Simias
3. Brachiation	(i) True brachiation.	Gibbon type of brachiation.	Hylobates, Symphalangus
	(ii) Modified brachiation.	Chimpanzee and orang-utan type of brachiation.	Gorilla, Pan, Pongo
4. Bipedalism		Striding	Homo

Locomotor classifications have some utility as an introduction to the variety and distribution of primate locomotor adaptations. The most commonly cited locomotor classifications are probably those of Napier and Napier (1967, Table 9-1) and Ashton and Oxnard (1964). Whereas these classifications do offer a description of the range of locomotor adaptations for primate groups, they also cause misconceptions. First, it is not clear if these are behavioral or anatomical classifications. Second, similar morphology has been interpreted to mean similiar locomotion; this relationship does not always hold. Third, similar locomotion has been interpreted to mean similar anatomical adaptation; this relationship also does not always hold. These problems become more evident in the following examples.

The classifications of Ashton and Oxnard (1964) and Napier and Napier (1967) recognize "semibrachiation" categories. Both include

Alouatta, Ateles, Brachyteles, and *Lagothrix* as New World "semibrachiators" and *Presbytis, Rhinopithecus, Nasalis,* and *Colobus* as Old World "semibrachiators," and in addition, Napier and Napier (1967) include *Pygathrix* and *Simias* in the latter group. A "semibrachiator," as considered by these researchers, is an animal that arm-swings fairly regularly, supplementing its locomotion with leaping and/or quadrupedal progression. This category may, however, be more cohesive morphologically then behaviorally. In musculature and skeleton all these animals have been found to be intermediate in form between "brachiators" and "quadrupeds" (Napier, 1961; Ashton and Oxnard, 1963). Apparently, from this similarity in anatomical structure, the animals were assumed to share similar locomotor adaptations. Recent field research has demonstrated that the animals grouped in this category have widely varied locomotor adaptations. At one extreme Mittermeier (1978) and Fleagle and Mittermeier (1980) found, in *Ateles paniscus,* arm-swinging accounting for 38.6 percent of their total locomotion, and in *Ateles geoffroyi* 26.0 percent. Locomotion of these two species also included over 20 percent quadrupedal locomotion and climbing. Leaping accounted for 11.4 percent (*A. geoffroyi*) and 4.2 percent (*A. paniscus*) of their locomotion. Napier and Napier (1967: 385) describe the locomotion of New World "semibrachiators" as "arm-swinging with use of prehensile tail; little leaping"; this is fairly apt for *Ateles. Alouatta's* locomotion, however, is reported to include very little, if any, arm swinging (Mendel, 1976; Fleagle and Mittermeier, 1980). Thus, semibrachiation does not describe the locomotion of *Alouatta.* Morbeck (1979) reports that suspensory locomotion accounts for less than 2 percent of the total locomotion for *Colobus guereza;* again semibrachiation does not describe this animal's locomotion. Finally, Fleagle (1977) reports that the locomotion of *Presbytes obscura* and Presbytes *melalphos* includes less than 5 percent arm-swinging. So of the four genera of "semibrachiators" for which we have qualified locomotor data, only one, *Ateles,* brachiates habitually. Although the animals included in the "semibrachiators" by Ashton and Oxnard (1964) and Napier and Napier (1967) may be fairly similar morphologically, they certainly are not a cohesive group in terms of locomotion.

Napier and Napier (1967) recognize vertical clinging and leaping (VCL) as a distinct locomotor category. They describe VCL as "vertical leaping in trees and hopping on the ground." Included in this category are *Avahi, Galago, Hapalemur, Lepilemur, Propithecus, Indri,* and *Tarsius.* Leaping may in fact be an important part of the locomotion of all species of these genera. The locomotor group, however, is artificial because it includes two different locomotor adaptations (Cartmill, 1972). Galagos and tarsiers both have a greatly elongated calcaneum and navicular in their foot. This is their primary structural adaptation for upright jumping. In contrast, the other members of the group have elongated thighs and metatarsals as their structural adaptation for upright jumping. One weakness is the concept of "VCL" then is that it groups together animals with distinctly different structural adaptations. Further, all of these animals do not habitually leap to and from vertical substrates. McArdle (1981) reports that although *Galago alleni* is primarily a "vertical clinger and leaper," *Galago senegalensis* and *Galago elegantalus* leap in between substrates of many orientations. Finally, McArdle (1981) states that although all species of *Galago* are capable leapers, *Galago crassicaudatus* and *Galago demidovii* are primarily quadrupedal runners and climbers. Thus, "VCL" includes animals whose locomotion can be correctly categorized as "vertical clinging and leaping" (e.g. *Galago alleni*, and species of *Tarsius* and Indriidae) and animals whose habitual locomotion cannot be correctly categorized as "vertical clinging and leaping" (e.g. most *Galago* species).

When one looks at a "locomotor" classification, it should first be determined if it is a morphological or behavioral classification. Second, it should be remembered that similar morphology does not necessarily equal similar locomotion and vice versa. Third, one should realize that the categories in such classifications are grossly defined and include variation (both morphologically and behaviorally). This third point needs to be considered during the following discussion of locomotor behavior of extant primates.

Whereas all primates are capable of a wide variety of locomotions, they can be characterized by one or two locomotions that they use most frequently. All of the prosimians are primarily arboreal, although *Lemur catta* spends up to one third of its time on the ground. Galagos and tarsiers share a similar foot morphology: an elongated calcaneum and navicular. This appears to be an adaptation to leaping, the primary mode of locomotion for these small nocturnal primates (Charles-Dominique, 1977) that are highly specialized. Members of Indriidae are also primarily leapers; however, they do not share the foot adaptation of galagos and tarsiers. Members of Lorisidae can also be categorized by distinctive locomotion. They are slow, cautious climbers, and the pottos and slow loris prefer to locomote on larger substrates than do the slender loris and angwantibo (McArdle, 1981). All the other prosimians have usually been lumped together as arboreal quadrupeds.

The New World monkeys are also highly arboreal animals. *Alouatta* is a slow deliberate climber, showing little acrobatic behavior (Mendel, 1976) and always using the tail when climbing. The primary locomotor adaptation of *Ateles* is suspensory locomoting, including frequent arm-swinging with the use of a third limb, the prehensile tail in the process (Mittermeier and Fleagle, 1976). When they are moving fast, leaping appears to be the primary mode of locomotion for some of the smaller New World monkeys such as *Pithecia*, some of the callitrichids like genus *Cebuella*, the smallest monkey. Other New World monkeys for which detailed data on locomotion have been collected include *Saguinus, Midas, Saimiri sciureus, Chiropotes satanas,* and *Cebus apella* and it has been reported that these animals are all primarily arboreal quadrupeds.

There is also a wide variety of locomotor adaptations for Old World monkeys. A number of these monkeys spend most of their time on the ground and are often called terrestrial quadrupeds. These include the baboons, geladas, patas monkeys, and some of the macaques. When these animals are on the ground, they do not walk on the palms of their hands, rather they walk on dorsi-flexed digits 2–5. This is called digitigrade walking; see Figure 6–21. Most of the other Old World monkeys spend the majority of their time in trees and can be considered acrobatic arboreal quadrupeds. Some of the Colobines, however, probably leap more than the other Old World monkeys. Forms that incorporate a lot of leaping in their locomotor repertoire include *Colobus guereza* (Morbeck et al., 1979 and *Presbytis melalophos* (Fleagle, 1977a and b).

The locomotor behavior of the apes was for a hundred years relatively little-studied, and the concept arose that all the apes were arm-swinging forms. From this, paleoanthropologists deduced that man's ancestors must also have been brachiators. We know today that only the lesser apes—gibbons and siamangs—can be said to be real brachiators. Fleagle also studied the locomotor behavior of siamangs and found that brachiation was the primary mode of locomotion. In contrast, Tuttle (1965) documented that the African great apes exhibit a unique type of locomotion very different from brachiation; he described this type as quadrupedal knuckle-walking. Even so, African apes do sometimes move around, especially when young, by means of arm-swinging. Yet this type of movement is not identical with the arm-swinging locomotion of the lesser apes. In the largest African ape—the gorilla—locomotion exclusively by arm-swinging almost never occurs. Especially in males, the body size is so great as to diminish most climbing activities. Brachiation, upright walking, and knuckle-walking of African great apes on their fore-fingers are all specialized locomotor types in their own right, and there is no reason to believe one or the other need necessarily have evolved from the third of them. Horn (1976) has observed the locomotor behavior of the pygmy chimpanzee. He saw the pygmy chimpanzee—the bonobo or *Pan paniscus*—locomoting both on the ground and in trees. Pygmy chimpanzees are primarily quadrupedal both in the trees and on the ground; however, although they knuckle-walk on the ground, they are primarily palmigrade walkers in trees. Chimpanzees and gorillas knuckle-walk by flexing fingers 2–5 and walking on the dorsal aspect of the first and second phalanges of three digits. Apparently, the pygmy chimpanzee leaps and arm-swings in the trees more than the common chimpanzee does. As already stated, locomotion of gorillas appears to be similar to the common chimpanzee except that gorillas are more terrestrial. Field studies of the great Asian ape—the orangutan—are also much more extensive now than only a few years ago. Here again, it became apparent that orangutan locomotion is also unlike that of the lesser apes. They move about most carefully and deliberately, using all four extremities and swing only rarely with their arms alone.

After careful examination it has become apparent that orangutans are probably the most arboreal of the four great ape species but, like the common chimpanzee and gorilla, they do not commonly progress through the trees by brachiation. African and Asian apes differ not only in their methods of locomotion but also in the morphology of their forearms, wrists, and hands. Inaccurate knowledge of actual locomotor behavior and erroneous correlation of an overall similiarity (broad trunks) to an assumed function (brachiation) have led to misleading explanations of morphological features shared by extant men and the apes, and consequently, to misinterpretation of fossils. Only recently have we understood that our knowledge of extant primate locomotion must be profound in order to provide a source for interpretation of the locomotor apparatus of extinct species.

Most workers in primate behavior have neglected study of feeding and locomotor behavior of wild primate species in favor of attention to social interactions of animals. A few earlier studies that did report feeding and locomotion are summarized in Napier and Napier (1967) and by Rose (1973, 1974). The long-term observations of gorillas (Schaller) and chimpanzees (Van Lawick-Goodall) have included some locomotor observations, but with these authors also, the main interest focused on activities other than locomotion. Nevertheless, it has been a widely used practice to classify all the primates into what we called "locomotor groups" (one of which we have already discussed here, namely brachiation). It is hardly surprising that the number of basic locomotor groups, their naming, and the attribution of given species to particular groups differ widely between various authors.

In addition, most students have found it difficult to define locomotor terms. The definition of a particular locomotor type is, however, more easily achieved in those primates where a real specialization of locomotion exists. When we look at the wide variety of locomotor activities in primates, there appear to be several basic types of differentiation. As already stated, all of the lower primates are primarily arboreal except *Lemur catta*, which shows some degree of terrestrial adaptation. We also know of one generalized and two highly specialized and different locomotor types among prosimians. Many of these lower primates move around with upright trunks, leaping from one vertical support to another, where they land and cling in an upright posture. This peculiar locomotion has been described and defined by Napier and Walker (1967) as a distinct locomotor category, "vertical clinging and leaping." This kind of locomotion is said to apply to some of the smallest as well as to some of the largest extant prosimians. It characterizes the Indriidae, *Indri, Propithecus*, and *Avahi* and is also the locomotor type of galagos and tarsiers.

Lorisidae (pottos and lorises) can also be classified in terms of locomotor similarities; they usually move comparatively slowly and climb about holding on with three extremities at a time. They are thus called "slow climbers," even though they can move very rapidly, for example, when catching prey. All the remainder of the lower primates have usually been lumped together as arboreal quadrupeds.

The New World monkeys are also arboreal animals. These creatures run and leap, actively using all four extremities for grasping. Species of the four larger genera of South American monkeys are prehensile-tailed, using their tails as an additional fifth limb during locomotion. They also can move around by propelling or swinging themselves with their arms from branch to branch but not in the same way as the gibbons. This swinging led to the classification of the spider, woolly, and howler monkeys as semibrachiators. *Semi* is a Latin word and means "half." The definition of such a category is not really satisfactory. How can a manner of locomotion be half-something? It would be better to call these monkeys prehensile-tailed quadrupeds because all the other non-prehensile-tailed South American monkeys are generally regarded as arboreal quadrupeds. Semibrachiation also fails to describe the locomotion of any Old World monkeys. Particularly among these, the colobines have been regarded as semibrachiators. Colobines use their arms when moving through the trees; but just as with the four larger cebids, colobines certainly do not brachiate in the sense of the locomotion of lesser apes.

Many of the Old World monkeys spend a considerable amount of their activity-time on the ground. Some of these—for example, *Erythrocebus patas* (the Patas monkey), the baboons, and many of the macaques—stay most of the time on the ground. Consequently, they have to be considered mainly terrestrial monkeys, in spite of the fact that they move back into the trees or onto steep rocks for the night. These monkeys differ enough in their locomotion from those monkeys that live principally in trees that they are classified as terrestrial quadrupeds.

We have already discussed in part the locomotion of the apes. The chimpanzee and the gorilla bear weight on the outside of the flexed digits of their hands and have dermatoglyphics on the skin pads on the dorsal aspect of these fingers. As we have seen, all apes have elongated forelimbs, a feature more evident in the lesser apes. These long forelimbs were regarded as indicating that all apes were arm-swingers. Later, Jolly (1972) demonstrated that forelimb elongation need not necessarily correlate with brachiation. The extinct giant open-country baboons he studied were too large for trees, lived in arid environments, and yet had long forelimbs. Some of these huge baboons were as large as a female gorilla, and at least one species had forelimbs longer than hindlimbs. They were, at first, ranked in a genus of their own, *Simopithecus*, but Jolly placed them in the same genus as the present-day gelada baboon, *Theropithecus*, a highly terrestrial primate that lives today in the treeless high country of Ethiopia. We know that *Theropithecus* certainly cannot brachiate.

Limb proportions have also been widely used as indicators of the locomotor type held by various primates. Thus, it has always been believed that long hindlimbs had to be correlated with vertical clinging and leaping, or at least to leaping locomotion. It is not correct to assume that because all vertical clingers and leapers have long hindlimbs, *ergo* a primate with long hindlimbs is a vertical clinger and leaper. "All brachiators have long forearms; all primates with long forearms are brachiators" falls in the same category of mistakes. Man himself has very long hindlimbs compared to his body length or the length of his forearms. The length from shoulder to wrist is about three quarters that from hip to ankle. These long legs are without doubt correlated to man's upright, bipedal gait.

When the proportions are this different, it is possible to see that limb length plays a part in the locomotor abilities of the animals concerned.

Most primate species do have the capacity for a wide variety of locomotor activity. Long legs result in long strides and the hands are freed for carrying: These make bipedal movement efficient. Long arms and a highly mobile shoulder articulation improve arm-swinging or brachial movement through the upper strata of trees. Elongated hindlimbs may be effective in leaping. We have seen that long arms can also be of advantage for primates that move quadrupedally on the ground (knuckle-walking) by bringing the head into a better position for seeing. Long arms are also useful in reaching after food objects for primates that sit a great deal and that typically remain seated upright when foraging (*Theropithecus gelada*). Macaques and baboons in one case and the great African apes in another functionally elongate their forearms by walking on digits—not palms—and thus extending the length of the arms. This is achieved in different ways in the two groups, but the effect is the same: In macaques and baboons the bottoms of digits 2–5 of the hands touch the ground, whereas carpals and metacarpals stay in line with the elongated forearm. Among knuckle walkers these same fingers 2–5 are flexed in plantar direction, and the animals walk on the inverted dorsal aspect of the first and second phalanges of these fingers. This also elongates the forearm by adding the length of the carpals as well as that of the metacarpals and the basal phalanges. All the specialities in limp length described above are correlated to a certain kind of specialization in the way these primates move about. This in turn shows us that special types of locomotion and specializations in the locomotor apparatus do go together. Only where locomotor function in primates is truly specialized do we also find a specialized morphology. In this context one observation is of interest: In humans, it appears that even with the combination of comparatively unspecialized hands, an extreme hindlimb adaptation, and a highly developed brain, a high degree of ability to adapt to a broad range of locomotor activities and to a great variety of environments is retained.

One possible approach to defining locomotion in primates is to measure the percentage of time spent during periods of activity in one or the other way. How much time does the average adult human being today spend moving about in our typical upright walk? Between 10 and 50 percent of the time during a ten-hour average day is said to cover the extremes. Equally important is what adult humans do not habitually do, but are, in fact, capable of doing, namely for example no percent leaping, swimming or crawling. It has also been estimated for today's true brachiators, the lesser apes, that they brachiate about 80 percent of the time when actively moving.

There is now a great deal of information documenting the locomotor behavior of extant primates. It is still difficult, however, to offer satisfactory hypotheses about the locomotor adaptations of fossil primates based strictly on comparisons with extant primates. A number of primates have similar morphological characteristics, yet their locomotion is quite different. For example, geladas, gibbons, and orangutans all have elongated forelimbs. The long arms of gibbons appear to be an arm-swinging adaptation, whereas the long arms of orangutans are used in powerful climbing or hanging while feeding. The long arms of extinct giant geladas were interpreted as an adaptation for reaching food items when in an upright seated position.

Before we analyze the post-cranial remains of fossil primates then, we need a better understanding of the relationships of such factors as environment, body size, and morphology with locomotion. The relationships of environment, body size, and locomotion can be analyzed in field studies, and the relationship between anatomy and locomotion can be best understood through analysis in the laboratory. A number of such studies have been carried out over the past few years. Summaries of some of these research projects will illustrate what can be learned from this type of analysis.

Fleagle's (1977a, 1977b) research on two sympatric species of Malaysian leaf monkeys, *Presbytis obscura* and *Presbytis melalophos*, provides valuable insight of possible correlations between locomotor behavior, substrate use, and anatomical structure. He found *P. obscura* to be primarily a quadruped monkey that prefers to locomote on large supports. In contrast, *P. melalophos* leaps much more frequently and prefers to locomote on smaller supports. Further, *P. obscura* spends a majority of its time in the horizontally continuous main canopy, whereas *P. melalophos* frequents the discontinuous underside of the forest. Finally, Fleagle has documented a number of statistically significant anatomical character differences between these two species. In each case these differences make "mechanical" sense when considered with the locomotor behaviors of these primates.

Mittermeier and Fleagle (1976) carefully compare and evaluate similarities and differences in the locomotor behavior of *Ateles geoffroyi* and *Colobus guereza* and spell out the uselessness of locomotor categories such as "semibrachiation." Both these monkeys have repeatedly been classified as "semibrachiators," but it turns out that their locomotor habits are as different as can be. Also, the authors caution that it is almost certainly misleading to generalize locomotor behavior from one particular population to an entire species. Troups that live in different environments often exhibit clearcut locomotor differences within the same species, differences that then are most likely not to be reflected in the morphology.

Another interesting field study is that of Fleagle and Mittermeier (1980), in which they document the locomotion of seven species of New World monkeys in relationship to these monkeys's body size and ecology. For the seven species studied—*Sanguinus midas, Saimiri sciureus, Pithecia pithecia, Chiropotes satanas, Cebus apella, Alouatta seniculus,* and *Ateles paniscus*—Fleagle and Mittermeier noted the following trends: Smaller forms tend to leap more than larger forms, whereas larger forms tend to climb more than smaller forms. Also, larger forms tend to use larger supports; however, this trend has two exceptions. *Sanguinus* preferred relatively larger, and *Ateles* preferred relatively smaller substrates than would be predicted from body size alone. Finally, these authors found no relationship between locomotor behavior and diet.

Experimental research on primate locomotion is very important because it provides a rigorous test for hypothesized morphology-behavioral relationships. Radiography and electromyography (EMG) are presently the most powerful experimental techniques for testing function. Jenkins (1981) recently identified by radiography a morphological adaptation for brachiation in the wrist of *Ateles* and *Hylobates*. He found that the proximal row of carpals form a socket that rotates around a ball formed by the capitate and hamate. Jenkins did not find this morphological complex in the quadrupedal primates he examined. Thus, this research has identified a morphology, which if found in a fossil form would be highly suggestive of a brachiation adaptation.

Stern and Susman (1981) examined by telemetered EMG the function of the *gluteus medius* muscle of *Hylobates lar, Pongo pygmaeus,* and *Pan troglodytes.* Most researchers had considered this muscle to be primarily an extensor of the thigh on the basis of its origin and insertion. Stern and Susman, however, found the *gluteus medius* muscle of these apes to function primarily as a medial rotator of the thigh, as it does for humans. These authors conclude that during bipedal walking the gluteus medius muscle provides side-to-side balance of the trunk at the hip in both humans and apes. This research is an example of how hypothesized morphological-behavioral relationship can be rigorously tested by an experimental technique. It had been accepted by most researchers that the gluteus medius muscle functioned as an extensor of the thigh in apes. Research using EMG now has documented that this muscle fires for these animals when they medially rotate the thigh, not during extension. The assumption by Stern and Susman, however, that "...*gluteus superficialis* need not have changed its action in the step from pongid to hominid" is misleading. There is no fossil pongid pelvis to show that the pelvic morphology of extant apes was already manifest in a pongid that could qualify as a forerunner for hominid evolution. Extant apes have very specialized pelves. In fact, the pelves are so specialized that they can hardly be considered as a possible stage during the evolution of the human pelvis/femur complex.

A relationship between locomotion and anatomy has been recognized for many years. Recent research suggests, however, that posture may also be a factor in the evolution of limb and trunk morphology. A number of field studies, including Morbeck et al. (1979), Mendel (1976), and Rose (1974) have found that primates spend less than 10 percent of the day locomoting. The vast majority of the time of these animals is spent in postural activities. A number of postures (hanging, clinging and sitting, for example) provide stress and strains on the anatomy of the animal. These postures are undoubtredly factors that help shape the anatomy of the animal. Posture is something that researchers need to consider when analyzing the anatomy of an animal to ascertain functional or adaptive specific morphological characters and complexes.

For additional information on primate locomotion and its relationships with anatomy and environment, see Jenkins (1974), Morbeck et al. (1979), and Stern and Oxnard (1973).

In the past ten years we have seen a vast increase in our knowledge of primate locomotion. There are still many primates, including prosimians, New and Old World monkeys, and apes for which we need information of locomotion. If and when these animals are studied, researchers need to consider when, where, and how often the animals locomote, and also, they should consider what postures are utilized by the animal.

chapter 10
Chromosomes and Blood Groups

CHROMOSOMES

Little more than twenty five years have passed since scholars learned that the number of chromosomes in most of the cells of our tissues is 46 and not, as had been thought before, 48. Perhaps the reason for the delayed discovery of this important fact about ourselves derived from the small size of primate chromosomes in general (including those of man) as well as the development of optical instruments with high powers of resolution and development of new staining and spreading procedures for chromosomes (Bender and Chu, 1963). Tjio and Levan (1956) and Ford and Hamerton (1956) were the ones to prove conclusively that the diploid number of human chromosomes is 46. This was the start of a major breakthrough in understanding chromosome morphology, especially in our species. It led in turn to a much better understanding of chromosome-linked heredity.

The word *chromosome* is derived from the Greek language and means "colored body." The name applies to those parts of each cell nucleus that are known to be the carriers of genetic information. These carriers are visible in certain phases of the cell cycle and become quite obvious and colored when tissues are stained in a certain way. Living cells go through different phases and continuously divide into new cells. The visibility of the chromosomes is highest when stained during the first half of mitosis, when the cells are beginning to divide at metaphase.

In this phase of highest visibility, each chromosome appears to be longitudinally double (i.e., consists of two identical elements), and each half is called a chromatid. Each chromatid generally consists of two arms linked together at a point called the *centromere*. The centromere may also be called the primary constriction because the breadth of the chromosome is highly reduced at this point. In functional terms the centromere is called the *kinetochore* (Greek for "center of movement") because it is the place of attachment for the apparatus that pulls the chromatids apart during division. Each chromatid is made up of the highly folded and packed molecular thread comprising a chain of the units of heredity, the genes. Single genes are not, however, visible on the chromosome. Genes and chromosomes are subject to constant alteration, genes by mutation and chromosome by breaking and rejoining. Because genes are intimately linked to the entire character of the organism, its form and its function, it follows that variations in the chromosome structure or number should provide some information on the relation, and possibly even the evolution, of animals. A practical difficulty is that mammalian and especially primate chromosomes are very small, and it is therefore difficult to work with them.

Basically, there are two types of cells in the mammalian body: the common cells of all the tissues of the body—the somatic cells—and the germ cells of the testis and ovary. Mitotis (cell division producing identical daughter cells) occurs in the somatic cells that have two similar sets of chromosomes (2n), one inherited from each parent (this means there is a pair of each chromosome present in these cells). The mature germ cells—the gametes and ova—have only one set of chromosomes (1n) because in undergoing meiosis during their development, the 2n set is reduced to half the full complement of chromosomes (one of each pair). For example, human germ cells—sperm and eggs—contain $1n = 23$, whereas most cells of the body tissues have $2n = 46$ chromosomes. There are, however, some exceptions to this rule; some organ tissues, the liver for example, contain multiple sets of chromosomes. The typical set of chromosomes of an animal species is called its *karyotype*. The karyotype of an animal is characterized by the number and by the form of the chromosomes. The form or morphology of a chromosome is determined primarily by the position of the centromere. There are also differences in the intensity of color of some chromosomes. Some chromosomes have uncolored areas near the end of one pair; these uncolored areas are called achromatic (or unstainable) zones and are the secondary constrictions of chromosomes as compared to the primary constriction, the centromere. At metaphase secondary constrictions are seen in pairs, and the primary constriction—the centromere—appears single. Distally, the secondary constrictions are regularly followed by normally staining chromosomal areas called satellites. The secondary constriction region usually contains ribosomal DNA and is known as the nucleolus organizer region. Chromosomes with the same arm length and same position of the centromere that lack achromatic zones and satellites cannot be distinguished within one karyotype except by banding.

Karyotypes are subject to change: as already stated, chromosomes break and rejoin in many different ways. Parts of chromosomes can be lost or transposed in the process of such events. If a large part of a chromosome is lost within a cell, or if the reunion of chromosome pieces prevents the chromosome from being separated in half or otherwise from functioning properly and from being distributed equally at division, one or both daughter cells may die and be lost from the tissue. In the case of a germ cell, the daughter cell may be lost from the population, or if it takes part in fertilization, an abnormal offspring may result. Such changes of chromosomes and thus of the karyotype are called "rearrangements." Only through the germ cells do such rearrangements have evolutionary effects. There is another type of change that works at the level of the genes. These changes are called "gene mutations" and are invisible. Gene mutations alter the molecular combination of the desoxyribonucletides, and in consequence, they change the genetic information and hence the character of certain gene products containing them.

Let us now return to the method of describing an animal's karyotype. Both body tissue cells and immature germ cells can be used to count and describe the chromosomes. This is more complicated with cells from the various somatic tissues of the body because there is the double number of chromosomes (2n) that must be sorted out. In immature germ cells the number of chromosomes is the same, but when the cells are closely paired, the number appears to be 1n. However, the cells are more difficult to obtain, and the chromosomes more difficult to see than those of other tissues. It is useful to evaluate both types of cells of an organism whenever possible.

In most chromosome studies, karyotypes are constructed from mitotic metaphase chromosomes. Because of the development of techniques in the 1950s for culturing tissues and cells *in vitro*, it is now relatively easy to culture lymphocytes obtained from blood samples. The growing cells are arrested in metaphase by the addition of colchicine; the chromosomes are separated by swelling with a hypotonic solution, fixed, spread on a slide, and stained with dyes like Giemsa or orcein. These techniques emphasize uniformly stained chromosomes and are referred to as prebanding methods. Such treatment provides information about the relative lengths of chromosomes, the position of each centromere, and the presence of satellites.

Analysis of a karyotype is based on a photograph from which individual chromosomes are cut and arranged in a particular sequence. The chromosomes are seriated according to their absolute length, beginning with the longest individuals, and according to the position of the centromere. If the centromere is positioned medially (length ratio of the arm between 1:1 and 1:1.9), the chromosome is called metacentric. When a chromosome has a ratio of arm lengths between 1:2 and 1:4.9 and has a centromere with a submedian position, it is called "submetacentric" chromosome. If the arm length ratio increases about 1:5, and the centromere position is subterminal, the chromosome is considered "acrocentric." It can also happen that the centromere position is terminal and the chromosome has only one pair of long arms; in this case the chromosome is called telocentric (Figure 10–1). Telocentric chromosomes are not known in primates and can therefore be disregarded in this discussion.

In the early 1970s it was demonstrated that fluorescent dyes such as quinacrine (Caspersson et al., 1970) produced bright and dull fluorescent bands along the length of the chromosome. Such banding patterns are unique for each chromosome of a species. Banding patterns can be produced by fluorescent dyes such as quinacrine or acridine orange or Giemsa after certain pretreatments.

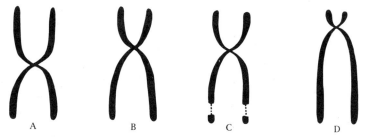

Figure 10–1 Diagram of the insertion types of different chromosomes: (A) metacentric chromosome (median insertion); (B) submetacentric chromosome (submedian insertion); (C) submetacentric chromosome with achromatic (unstainable) portions and satellites; (D) acrocentric chromosome (subterminal insertions).

The Q-banding pattern produced by quinacrine is the same as the G-banding pattern produced by Giemsa with a trypsin pretreatment. Reverse or R-bands are produced with either acridine orange or Giemsa after heat treatment.

Finer detail of the banding patterns can be achieved by making the chromosomes longer and using more recently developed banding techniques such as hi-resolution Giemsa banding (Yunis et al., 1978). Cell synchronization is used to obtain chromosomes at early metaphase, prometaphase, and even late prophase, when the chromosomes are longer and exhibit more bands. At prometaphase and later prophase approximately 1200 bands can be seen per haploid set in the human, whereas only 300–500 are seen at metaphase. With such newly developed techniques it is now possible to follow subbands not resolved on metaphase chromosomes and to look more closely at homologies that are thought to exist between species.

During prophase chromosomes become highly spiralized, shortened, concentrated, and thus visible. Chromosomes are transformed from a functioning dispersed chromosome into a discrete transport chromosome. The following events are possible and result in changes of the chromosome complement of cells, or mutation:

1. Entire sets of chromosomes (one set of chromosomes is equivalent to the *genome* of an organism) can duplicate without a subsequent division of their cell. This duplication results in an increase of the number of chromosomes (x times n), and the cell becomes *polyploid* (polyploid cells are present in a certain percentage in highly active organs, for example, the liver).

2. A change of single chromosomes can be achieved in different ways: one or more chromosomes may break, or pieces of chromosomes can be lost (deletion) or can be duplicated by addition or interchange of pieces (duplication); pieces can turn over and fuse again with the same chromosome (inversion) or can be exchanged between chromosomes (translocation). All these are termed "rearrangements." Many rearrangements are not able to survive and are lethal for the cells involved (Figure 10–2).

Rearrangements can naturally result in loss of some chromosome material (DNA). In mammals rearrangements can also result in a change of the number and/or the type of chromosomes of given karyotypes. However, the amount of DNA in each genome is roughly constant, namely 3.5×10^{-9} mg DNA in many mammals

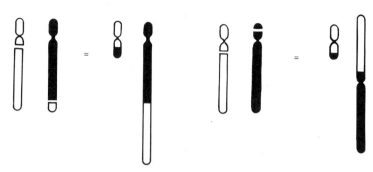

Figure 10-2 Chromosome mutations. Possible ways of recombination of two acrocentric chromosomes in two metacentric chromosomes.

(Ohno, 1969). It seems that most successful changes of genetic information cannot be achieved with less genetic material (DNA). The genetic material undergoes reorganization to produce evolution. Change of the genetic material is inherited and may lead to species differences; these in turn may lead to species separation, a step in evolutionary progression.

Some students of primate chromosomes have tried to detect differences in the amount of DNA per cell in different primate genera and species. The tiny differences found do not, however, exceed the range of error of the delicate procedures for measuring DNA content. For example, the data show that the amount of DNA in a diploid cell of *Lepilemur* and *Callicebus* with 2n = 20 in certain species (the lowest number of chromosomes within the order) and the amount of DNA of a *Tarsius* cell with 2n = 80 (the highest number of chromosomes within the order known until now) is the same and equals 7.0×10^{-9} mg. Although karyotypes differ (between different morphological species), they can also be alike.

Another disorder is when chromosome pairs divide disproportionately. For humans it is known, for example, that one pair of chromosomes can have a third equivalent. This equivalent is called "trisomy" and results in physiological and developmental deficiencies of the carrying individual. If a cell is deficient in a chromosome, it is "monosomic." Cells that have one or more whole chromosomes absent from or added to the normal complement (referred to as "euploid") are called "aneuploid." Aneuploidy may result from a failure of chromosomes to separate or disjoin properly during cell division, a phenomenon called "nondisjunction."

As already said, chromosomes of the diploid cell appear in homologous pairs, one member of which came from the mother

Table 10-1 Chromosomes of Prosimii and Tree Shrews

	2N	M	S	A	X	Y	References
Tupaidae							
Tupaia glis	60	8	4	46	S	A	Chu & Bender, 1961; Klinger, 1963*
	62	8	4	48	S	A	"
Lemuridae							
Lemur catta	56	4	4	46	M	A	Hamilton & Buettner-Janusch, 1977
Lemur mongoz	60	0	4	54	A	A	"
Lemur fulvus	60	0	4	54	A	A	"
Lemur macaco	44	12	8	22	A	A	"
Varecia variegatus	46	16	2	26	S	A	"
Hapalemur griseus griseus	54	4	6	42	A	A	Rumpler & Albignac, 1973*
Hapalemur griseus olivaceus	58	2	4	50	A	A	"
Lepilemur ruficaudatus	20	2	16	0	M	A	Rumpler, 1975*
Lepilemur mustelinus	34	0	6	26	S	A	"

Species						Reference
Microcebus murinus	66	0	0	S	A	Rumpler & Dutrillaux, 1979
Cheirogaleus major	66	0	0	S	A	"
Cheirogaleus medius	66	0	0	S	A	Dresser & Hamilton, 1979
Propithecus verreauxi coquereli	48	14	16	S	A	Rumpler, 1975*; Poorman, ms.
Lorisidae						
Nycticebus coucang	50	22	26	S	M	deBoer, 1973b*; Garcia et al., 1978
Perodicticus potto	62	16	28	S	A	deBoer, 1973b*
Galago senegalensis braccatus	36	30	4	A	A	deBoer, 1973a*
	37	29	5	A	A	"
	38	28	6	A	A	"
Galago crassicaudatus monteiri	62	6	48	S	A	deBoer, 1973a*; Poorman, in prep.
Galago garnetti	62	6	34	S	A	"
Tarsiidae						
Tarsius bancanus	80	(14	66)	?	?	Egozcue, 1969*

*Numbers based on unbanded karyotypes; karyotype description may change with banded chromosome analysis.

organism, the other from the father. In females there is one pair of sex chromosomes (the X chromosomes). In males the sex chromosomes, consisting of an X and a Y chromosome, are nonhomologous and differ in size and banding properties. Chromosomes other than the X and Y are called autosomes. Sex chromosomes are not easily differentiated from the autosomes in females. Thus, if only females of mammals have been investigated with regard to karyotype, the sex chromosomes cannot be identified because it is impossible to determine which pair of chromosomes are the two X's and which are the autosomes. The Y chromosome seems not to be of much evolutionary importance as a carrier or prime genetic information other than the determination of the male sex. Very small chromosomes that cannot easily be classified according to the position of the centromere are called microchromosomes. Chromosomes with very long arms are sometimes called macrochromosomes. Chromosome polymorphism is known to occur in mammals. This polymorphism is the occurrence of one or several chromosomes in two or more alternate structural forms within a population. Such populations are often geographically separated but are not regarded as distinct subspecies. In spite of the fact that we already know many primate karyotypes, multiple uncertainties are still involved; in many cases only a few specimens have been studied. Therefore, one can see why differences of karyotypes recorded for the same primate species are produced by different laboratory studies.

PROSIMII AND TUPAIIDAE

The number of chromosomes varies between 20–80 within the Prosimii. *Tupaia glis* is reported to have $2n = 62$ and $2n = 60$. The form with $2n = 62$ is missing one pair of large submetacentric chromosomes but has two extra pairs of acrocentrics not found in the variety of *Tupaia glis* with $2n = 60$. Two acrocentric chromosomes can fuse and form one biarmed (metacentric or submetacentric) chromosome (centric fusion). Thereafter, of course, the small acrocentric chromosomes are gone. Within the genus *Lemur* the variation in chromosome number ranges from $2n = 44$ to $2n = 60$. The number of chromosomes typically seen in *Lemur fulvus* subspecies is $2n = 60, 52, 51, 50,$ and 48. $2n = 58$ has also been reported for several individuals. Some species of *Lepilemur* have only 20 chromosomes (see Petter et al., 1977). The high degree of variation in the chromosome number of prosimians is accompanied by differences of length. Chromosomes belonging to species of the genus *Lemur* vary in length from very large macrochromosomes to very small microchromosomes.

A comparison of karyotypes within the prosimians reveals some regularity. If the 2n number of chromosomes is high, many chromosomes are acrocentric and few metracentric. If the 2n number of chromosomes is low, the majority are metacentric, and only a few acrocentric chromosomes can be found. In the transitional field between those extremes, namely around 2n = 50, lies the chromosome number of *Nycticebus coucang* 2n = 50, and it makes an exception to this rule by having only metacentric chromosomes.

This regularity of high chromosome number and a high proportion of acrocentric chromosomes in prosimian karyotypes lead to an evolutionary theory that will soon be discussed.

ANTHROPOIDEA: CEBOIDEA, CALLITRICHIDAE

The Callitrichidae are quite uniform in their karyotypes; they are also most interesting in relation to karyotype evolution. All the karyotypes of species of the family that are known have 2n = 44, 46, or 48. Most have 4 metacentric chromosomes, 24–28 submetacentrics, and 10–18 acrocentrics in their diploid set. The reduction of the diploid numbers from 48 to 46 and from 46 to 44 can be achieved by reducing the number of acrocentric chromosomes by two pairs and increasing the submetacentric number by one pair of chromosomes (for example, species of *Callimico, Saguinus, Cebuella,* and *Callithrix*). The mechanism in question is centric fusion: a large pair and a small pair of acrocentric chromosomes fuse to form one large pair of submetacentric chromosomes. The differences of the karyotype between *Saguinus fuscicollis* on the one hand and *Callithrix jacchus* and *Leontideus illigeri* on the other have been explained as the results of translocations of an acrocentric and a submetacentric pair of chromosomes producing two submetacentric chromosome pairs. Because most of the karyotypes of these species predate banding methods, chromosome banding is needed for substantiation of the hypothesis of centric fusion and rearrangement by translocation in these species. Banding studies would show whether homology exists between the arms of the submetacentrics and the autosomes from which they purportedly derive.

Callimico is a genus of the Ceboidea that has been much

Table 10–2 Chromosomes of Anthropoidea

	2N	M	S	A	X	Y	Reference
Callitrichidae							
Callimico goeldii	48	4	24	18	S	A	Egozcue et al., 1968[*]
Saguinus fusciollis illigeri	46	4	26	14	S	M	"
Callithrix jacchus	46	4	28	12	S	A	"
Cebuella pygmaea	44	4	28	10	S	A	"
Callithrix aurita	44	4	28	10	S	A	"
Callithrix argentata	44	4	28	10	S	M	"
Callithrix chrysoleucos	46	4	26	14	S	M	"
Leontideus rosalia	46	4	28	12	S	M	"
Leontocebus illigeri	46	4	30	10	S	M	Chu & Bender, 1961[*]
Cebidae							
Aotus trivirgatus trivirgatus	50	8	18	22	M	M	Yunis et al., 1977
Aotus trivirgatus griseimembra	54	10	10	32	M	A	Miller et al., 1977
	53	11	10	30	M	A	"
	52	12	10	28	M	A	"
Callicebus moloch cupreus	50	12	6	26	S	A	Benirschke & Bogart, 1976
Callicebus torquatus torquatus	20	2	6	10	M	?	
Pithecia pithecia pithecia	48	10	8	28	M	M	deBoer, 1975[*]
Cacajao rubicundus	45	10	8	24	?	?	deBoer, 1975[*]

Species	2n						Reference
Alouatta seniculus seniculus	43-45/44	10	6	26	A	S	Yunis et al., 1976; Egozcue, 1969*
Alouatta caraya	52	4	16	30	S?	A?	"
Saimiri sciureus	44	10	20	12	S	A	Lau & Arrighi, 1976, Ma et al., 1974
Ateles paniscus chamek	34	12	18	2	S	S	Kunkel et al., 1980
Ateles belzebuth belzebuth	34	10	20	2	S	A	"
Ateles geoffroyi	34	12	18	2	S	S	"
Lagothrix lagotricha	62	8	22	30	S	A	Dutrillaux et al., 1980
Cebus capucinus	54	6	10	36	M	S	deCaballer et al., 1976
Cebus apella	54	4	16	32	M	S	"
Cebus albifrons	54	6	12	34	M	S	"
Cercopithecidae							
Macaca mulatta	42	14	26	0	S	A	Dutrillaux et al., 1979
Macaca fascicularis	42	14	26	0	S	A	"
Cercopithecus mona	66	(48)		16	S	S/A	Egozcue, 1969*;
Cercopithecus mitis	72	(48)		22	S	S/A	Petter & Albignac,
Cercopithecus aethiops	60	14	22	22	S	A	Dutrillaux et al., 1978b
Papio anubis	42	14	26	0	S	A	Dutrillaux et al., 1978a, 1979
Papio papio	42	14	26	0	S	A	"
Papio hamadryas	42	14	26	0	S	A	Bernstein et al., 1980
Papio ursinus	42	14	26	0	S	A	"
Erythrocebus patas	54	16	28	8	M	?	Dutrillaux et al., 1978b
Colobus polykomus	44	(42)		0	S/M	A?	Egozcue, 1969*; Petter & Albignac

Table 10-2 (continued)

Pongidae	2N	M	S	A	X	Y	References
Symphalangus syndactylus	50	(46)		2	S?	S?	Egozcue, 1969*
Hylobates moloch	44	28	14	0	S	A	Tantravahi et al., 1975
Hylobates lar	44	28	14	0	S	A	"
Hylobates concolor	52	32	12	6	M	A	"
Pongo pymaeus	48	6	20	20	S	A	Paris Conf. Suppl., 1975
Gorilla gorilla	48	16	18	12	S	A	"
Pan troglodytes	48	10	24	12	S	A	"
Pan paniscus	48	10	26	10	S	A	Bogart & Benirschke, 1977
Homo sapiens sapiens	46	12	22	10	S	A	Paris Conf. Suppl., 1975

*Numbers based on unbanded karyotypes; karyotype description may change with banded chromosome analysis.

discussed with regard to its systematic placement. The dental formula of *Callimico* and the morphology of its skull are like those of the Callitrichidae. Part of the postcranial morphology and many behavioral traits resemble callitrichid monkeys rather than Cebidae. One of the karyotypes that has been recorded for *Callimico* also resembles callitrichids by having two pairs of metacentric chromosomes and a submetacentric X-chromosome. This karyotype of *Callimico* has been taken as an additional evidence for the taxonomic placement of *Callimico* in the family Callitrichidae. However, the fact that many of these karyotypes predate banding methods, and that these karyotypes have been published by different students should urge care in making such taxonomic placements until more data become available.

An XY and apparent XO condition is known in *Callimico goeldii*. However, in the "XO" male the Y chromosome is actually present, attached to a small acrocentric chromosome, and elongating one arm.

Cebidae

The number of diploid chromosomes of this family varies between 20 and 62 and thus shows approximately the same range of chromosome number as in the prosimians, which vary between 20 and 80. The species *Callicebus torquatus* has, together with species of *Lepilemur*, the lowest chromosome number of all karyotypes known in primates. *Lagothrix* has the highest number of chromosomes for cebid monkeys, $2n = 62$. Four species of *Ateles* have similar karyotypes with $2n = 34$. However, a large amount of chromosome variation exists between species and possibly between individuals within a species. Such chromosome variation may involve differentiation of populations. Three species of genus *Cebus* have $2n = 54$, but the two species of *Alouatta* differ in diploid number and karyotype ($2n = 44$ and 52). Two species of the rare family Pithecinae have the same chromosome number ($2n = 48$). The karyotype of the night monkey *Aotus trivirgatus* is highly polytypic, having a diploid number of 50 in *A. t. trivirgatus* and 52, 53, and 54 in *A. t. griseimembra*. The karyotypes of these animals show that hybrids exist between animals with different diploid numbers; for example, individuals with $2n = 53$ are hybrids between $2N = 52$ and $2n = 54$ individuals. *Saimiri sciureus* has at least 3 karyotypic races, each found in different geographical locations. The 3 races may coincide with 3 of the 7 subspecies of *S.*

sciureus. Individuals studied have 2n = 44 and may have 10, 12, or 14 acrocentric chromosomes with corresponding changes in the number of submetacentrics (12, 10, 8, respectively). Such changes between the number of acrocentric and submetacentric chromosomes, with the diploid number remaining constant, may result from pericentric inversions (inversions of chromosome segment involving the centromere).

Cercopithecoidea

Many karyotypes have been reported for the superfamily Cercopithecoidea. All the karyotypes known for the genera *Macaca, Cynopithecus, Papio, Theropithecus,* and *Cercocebus* have the same diploid chromosome number, which is 2n = 42. In addition, the karyotypes are identical in the total length of chromosome arms, consisting of all metacentric and submetacentric autosomes. The Y chromosome is tiny, the X chromosome submetacentric, and one pair of chromosomes is marked. Contrary to this, species of the genus *Cercopithecus* range in chromosome number from 2n = 48 to 2n = 72. The Y chromosome is a small acrocentric, the X chromosome submetacentric, and one pair of acrocentric chromosomes is marked. All species of *Cercopithecus* studied have all three types of chromosomes (metacentric, submetacentric, and acrocentric), the numbers of which are more or less equally balanced. *Erythrocebus patas* has a diploid number of 2n = 54. *Pygathrix nemaeus* has 2n = 44, and *Nasalis larvatus* has 2n = 48 chromosomes. The diploid number of *Presbytis entellus* is 2n = 44 and that of *P. obscurus* 2n = 44. *Colobus polykomos* also has 2n = 44. The total length of all chromosomes is not different in these species with different numbers of chromosomes. All the colobines already named have one pair of marked chromosomes that seems to be morphologically identical within the colobines.

Hominoidea

Hylobatidae

The gibbons *Hylobates lar* and *H. hoolock* have a diploid chromosome number of 2n = 44. Gibbon species have mostly metacentric, submetacentric, and few or no acrocentric chromosomes in their karyotypes. The Y chromosome of *Hylobates lar* is very small-shaped like a dot. For instance, in *H. moloch* the Y is somewhat larger and maybe metacentric. The *H. lar* Y looks like a dot. The other type of lesser ape, *Symphalangus*, differs clearly in its karyotype from that of *Hylobates*. The diploid number of *Symphalangus* chromosomes is 2n = 50; it lacks the pair of marked chromosomes and has one pair of acrocentric chromosomes.

Pongidae

The three pongid genera (*Pan, Pongo,* and *Gorilla*) all have the same diploid number of chromosomes, namely 2n = 48. The karyotype of *Pongo pygmaeus* is composed of 26 metacentric and submetacentric chromosomes and 20 acrocentric chromosomes, 16 of which are large and 4 small. 14 of the acrocentric chromosomes of *P. pygmaeus* have satellites. *Gorilla gorilla* has 34 metacentric and submetacentric chromosomes and 12 acrocentric chromosomes that also have satellites on their short arms. *Pan troglodytes* combines 34 metacentric and submetacentric chromosomes with 12 acrocentrics, 10 of which have satellites on their short arms. The karyotype of *Pan paniscus* resembles that of *Pan troglodytes*. One pair of short acrocentric chromosomes that are present in *P. troglodytes* is missing in *P. paniscus*, which have one more pair of submetacentric chromosomes than do *P. troglodytes*. In these four species the X chromosome is a large submetacentric; the Y is acrocentric and is one of the smallest chromosomes in the karyotype.

Hominidae

Man has the diploid number of 2n = 46 chromosomes, 34 of which are metacentric or submetacentric and 10 acrocentric. All these acrocentric chromosomes generally have satellites on their short arms. The Y chromosome of man is a small acrocentric and is somewhat variable in length. The X chromosome is submetacentric. The human karyotype is the best known of all primates, and many deviations from the normal karyotype have been detected. Furthermore, many of the effects of such karyotypological differences on the function and morphology of the human organism are known.

Detailed comparisons have been made of the chromosomes of man with those of the higher apes, using the banding methods discussed at the beginning of this chapter. Chromosomes that have identical banding patterns are described as homologous. A similarity in the banding pattern between different species has been shown to generally reflect a similarity in gene content. For example, counterparts for each chromosome of the human complement have been suggested in the chimpanzee, gorilla, and orangutan. The most conspicuous structural differences between the species can be explained by a series of pericentric inversions and differences in the amounts of heterochromatin (chromatin in a condensed state) at the centromeres and telomeres (noncentromeric ends of the chromosome). The presence of 46 chromosomes in man and 48 in the chimpanzee, gorilla, and orangutan can be explained by fusion of 2 acrocentric chromosomes to form the human chromosome #2. Evidence for the virtual homology of the chromosome bands of man, chimpanzee, gorilla, and orangutan comes from the localization of many genes to homologous chromosomes of the different species (Pearson et al., 1977; Garver et al., 1977); from the similarity of the amino acid sequences of a large number of proteins studied in each species (King and Wilson, 1975); and from a large degree of homology between repeated DNAs that are known to be transcribed (Gasden et al., 1977). The presence of identical banding patterns and gene contents in such widely divergent species indicates a conservation of linked genes with chromosome banding over tens of millions of years.

An evolutionary theory for the primate karyotype has been proposed. The primary assumption is that the number of chromosomes is more likely to be reduced than to increase. Many prosimians (and some other primates also) have a high number of acrocentric chromosomes when they have a high diploid number of chromosomes and vice versa. This condition has led to the idea that fusion of acrocentric chromosomes to form metacentrics with a decrease of the total chromosome number is an important mechanism in primate karyotype evolution. Evidence that karyotype rearrangement in the genus *Lemur*, for example, has taken place primarily by centric fusion of acrocentric chromosomes to yield metacentrics has been provided by chromosome banding studies. Banding analysis demonstrated the homology of the arms of most of the metacentrics in species that have lower diploid numbers with the acrocentrics in species that have higher numbers (see, for example, Rumpler and Dutrillaux, 1976; Hamilton and Buettner-Janusch, 1977). If fusion of a high number of original acrocentric chromosomes led to the many metacentric and smaller total chromosome numbers in higher primates, the prosimian *Nycticebus coucang* presents a problem because all its chromosomes are metacentric or submetacentric.

Another general assumption, which has been made, that primates with a high number of chromosomes are less advanced than primates with low diploid counts does not always fit with the taxonomic picture derived from the study of gross morphology in primates. *Tarsius*, for example, is by no means a generalized primate but *is* an extremely specialized prosimian, with the highest chromosome number (80) of any primate.

Explanations of the possible course of karyotype evolution within Callitrichidae are very interesting and even seem convincing. But these attempts have not explained the karyological differences within the genus *Cercopithecus* and other such phenomena. Basically, chromosome morphology compares groups and describes features of the chromosomes whose special contents and actions are unknown. As regards most primate characteristics, we do not even know on which chromosome the genes of such characteristics are located, much less the gene position on a certain chromosome. Chromosome morphology can be informative if evaluated together with numerous other characteristics. In humans gene loci are known for proteins or enzymes involved in malfunctions of the organism. That is, this knowledge has been derived from the study of deviations from the norm.

Hybridization between different species producing fertile offspring is of interest in the study of karyotypes. It was believed that species with similar karyotypes would be most likely to produce fertile offspring. Contrary to this theory, species with different karyotypes (*Cercopithecus stulhmanni* and *Cercopithecus ascanius*) have produced fertile offspring, whereas others with very similar karyotypes failed. Embryonic development involves an ordered expression of many genes that were inactive in the zygote. If the two genomes in an interspecific zygote are regulated similarly, an orderly development of the hybrid organism can be expected. However, if the patterns of gene expression are different, the probability is low that an interspecific zygote will develop successfully. In some of the cases where interspecific crosses had karyotypes that seemed identical, the offspring of the first generation were either abnormal and died early or, if they lived to become adults, proved unable to produce a second generation. Other species with identical karyotypes (as between species of *Macaca* and *Papio*) do produce fertile offspring. Thus, it has not yet been possible to learn much from hybridization in primates.

More knowledge, at least for evolutionary speculations, has been derived from chromosome chimerism known to ocur in Callitrichidae. The fact that Callitrichidae usually give birth to twins (or even triplets) might be one of the reasons for the occurrence of individuals with so-called *cell chimerism*, (Ford and Evans, 1977), namely various cells that differ in their karyotypes within individuals. Cell Chimerism is found only in Callitrichidae with an opposite-sex twin, and the chimerism affects the sex chromosomes. Consequently, a certain percentage of the cells in one individual are male and the others are female. The occurrence of chimerism has been explained by cell exchange in very early foetal stages of heterosexual dizygotic twins through the anastomoses of blood vessels in connected chorions. A pair of autosomes can also be heterozygous in *Callithrix* species. The existence of such heteromorphic autosomal pairs in living individuals of Callitrichidae suggests that the two different autosome types do not affect the individual's ability to produce offspring.

Biochemical Evolution and Molecular Clocks

Biochemical differences between animals were used to determine taxonomic distance or affinity soon after the turn of the century and increasingly in the 1920s. The increasing precision of biochemical techniques during recent years has resulted in the accumulation of much information in this field. With knowledge of the structure of DNA (the fundamental genetic material), Watson and Crick (1953) brought about one of the most recent revolutions in biology and biological thinking. An "era of analysis" was begun and soon provided data for primatology: a veritable mountain of recent publications on human and primate biochemistry. This situation has also led to detection and cure in man of many illnesses connected with biochemical disorders. In addition, primate taxonomy was approached with a new perspective. A significant field of scientific work concentrates, for example, on serological differences between various human populations. There can be no doubt that the number of human beings whose serology has now been studied amounts to millions living all over the world, even in remote areas. Such comprehensive knowledge, however, has been impossible to gather for nonhuman primates to date. The information is restricted by the relatively small number of individuals from different species that are available for such studies. Most primates living in captivity are of unknown geographical origin, and homogeneous captive populations are rare. Consequently, we cannot expect valid immunological information on primate "races." Nevertheless, some biochemists working with primates have, in spite of these limitations in assessing the actual composition of "primate samples," interpreted such artificial groups of apes and monkeys as "natural populations," which naturally led to erroneous results. So far, intrageneric biochemical taxonomy of primates has not been of much use. Intergeneric information, however, has some use as long as one is cautious and treats biochemical characteristics as of no more than equal value with numerous other morphological characteristics. The same caution is true for the evaluation of karyotypes.

Biologists since Darwin have accepted the evolution of plant and animal life. The only documents for evolutionary changes of life forms through time are fossils. Fossils of primates primarily consist of teeth, jaws, parts of skeletons, and sometimes endocranial braincasts. The biochemical tests that can be used on living organisms cannot be applied to fossils, yet fossils are the actual documents from former geological times. A principal feature of evolution is that different parts of organisms evolve through time at different rates. This phenomenon is called mosaic evolution. Yet another fact must be taken into account: When assessing biochemical similarity, structures that are alike and function similarly in animals can be, but are not necessarily, of the same origin in taxonomically related organisms, i.e., the structures can also be the result of parallel evolution. Thus, these (and other) facts long known to morphologists must also be adopted by people studying biochemical evolution. As Remane (1967) has pointed out, two species that have been separated for a long period of time can still be quite similar, whereas a third species that has separated recently can have changed considerably in a short stretch of time; see also the Appendix.

In addition to other factors, it appears that the speed of evolution can be affected by generation length. Lovejoy (1972) discusses generation length estimates that can be worked out for recent primates but can only be guessed at for those of past geologic times. Here again, fossils can be of help, provided they have been found in sufficiently large numbers. It is very unlikely that average generation length has been the same throughout the whole seventy or eighty million years of primate history. Sarich and Wilson (1968) have proposed the application of time values to immunological distances between homologous serum proteins (albumins) of recent primates. The rationale for immunological investigations is that the degree of biochemical difference is equal to the phylogenetic distance separating two or more species compared. In these investigations animals such as rabbits are injected with proteins from a donor species. These foreign proteins act as immunizing antigens. If the configurations of amino acids at the surfaces of the donor proteins differ from those of the host animal, antibodies are produced. Tests are then carried out to compare the proteins of the two species by observing the amount of cross-reaction, basically the amount of antibody-antigen precipitate. As the surfaces or antigenic properties of the proteins are altered by amino acid substitutions over evolutionary time, the extent of such cross-reaction decreases with increasing evolutionary separation between species. These tests therefore provide a way to measure phylogenetic distances.

Belief in molecular clocks is based on an assumed constant rate of accumulation of amino acid substitutions in proteins. Given constant rates of accumulation, an estimate can be made of the time elapsed since a pair of present-day species had a common ancestor (Wilson et al., 1977). A phylogenetic tree has been constructed with branch points providing estimates of the times of divergence of major primate groups. Among the many other discrepancies the branching events that separated the human lineage from that leading to the apes is much later than that determined from fossils dated geo-chemically. Even so, exponents of the "clock" believe that this method is better than determining primate history from fossils dated geo-chemically. Their "clock" was calibrated by assuming the evolutionary branch point of the Anthropoidea–Prosimii at sixty million years (Sarich and Wilson, 1968). Actually, this date is likely in error, the correct one should perhaps be estimated as closer to eighty million years ago, when basal placental groups diverged. At least five different families of primates had already arisen by sixty-five million years ago. It would not be unreasonable to suppose that such a degree of taxonomic diversity had taken fifteen million years to produce. The time of separation between the major groups of primates have been estimated by Sarich (according to serological distances) as follows:[1]

Homo – African apes	3.5 ± 1.5
Homo and African apes – Orangutan	7 ± 1

[1]Although there are many uncertainties in evaluating the basic branch points among primates from their fossil record, many groups resembling modern families seem to have diverged earlier than the late split point times favored by Sarich, Wilson, and their followers. A catarrhine-platyrrhine split as late as 36 MY comes long after north and south Atlantic rifting made transAtlantic rafting impossibly long in duration. Pottos and bush babies are distinct already at twenty million years in east African Miocene deposits, as are a variety of great and small apes. Those who favor a relationship between *Sivapithecus* and the orangutan have difficulties explaining the *Sivapithecus* appearance date, more than 14 MY. *Sivapithecus* appears much earlier than these apes could have been a distinct group that was similar to orangutans.

Pottos and Bushbabies	10 ± 1
Cercopithecoidea – Hominoidea	22 ± 2
Catarrhine – Platyrrhini	36 ± 3
Anthropoidea – Prosimii assumed	60

This immunological "clock" is neither more justifiable nor more correct than any other biological "clock" could be. But for the clock to even keep time, it is necessary to assume that rates of biochemical change are constant. There is nothing wrong with biochemical distance data in and of itself. The discrepancy between the "molecular clock" and paleontological data may be due to the first primate split-point being estimated at too low a number (sixty million years) as well as to variations in rates of evolution in different lineages.

An evolutionary clock has also been used to estimate rates of evolutionary change in karyotypes by considering the number of chromosomes and number of chromosomal arms per diploid genome. Evolutionary changes may be brought about by chromosomal rearrangements that affect the arrangement of large blocks of genes. Wilson et al. (1977) conclude that a correlation exists between the rate of karyotypic evolution and the rate of phenotypic evolution.

It seems fair to conclude then, that despite the great amount of effort devoted in recent years to the study of nucleic acid and protein evolution, serious problems arise when one tries to reconcile the dates of organismal evolution with those determined from macromolecular studies.

BLOOD GROUPS

As early as 1925 the first thorough studies of the blood of apes and monkeys were published by Landsteiner and Wiener. Molecular biologists have produced a great number of publications on major blood groups in primates and also on more rare and intricate blood substances since then. Interestingly, a number of otherwise comprehensive books on primates avoid this subject and plainly ignore the topic of primate blood groups. We will not go into great detail here either, especially because the information available still does not cover the broad range of primate species. Rather the information focuses mainly on such species as are readily available for laboratory research. Recently, some species have also been sampled in the natural environment. Even so, blood samples of primates can not be obtained as easily as those of humans, and even humans occasionally cause problems in such matters (for details on primate blood groups, compare Erskine and Socha, 1978).

Two techniques are used to test blood groups in nonhuman primates (Wiener, 1970). 1. Primate blood can be tested for human-type blood factors. In such tests reagents are used that have been originally prepared for typing blood of humans. 2. Primate blood can be tested with reagents that are obtained either by immunization of experimental nonprimates or preferably by iso- or cross-immunization with the blood of monkeys and apes. Blood types that are recognized by either of these procedures are accordingly called human-type or simian-type blood groups. The distinction between the human-type and simian-type blood groups is, however, somewhat indistinct.

The well-known A B O blood groups have been extensively studied in humans, apes, baboons, macaques, a couple of other Old World monkeys, and several New World monkeys; see Table 10–3. It appears that the blood groups are shared among many of the primates. A number of lemurs have B-like antigens on their erythrocytes. In monkeys, A B O antigens can be detected on some tissue cells and in the saliva (Socha, 1980). It has been stated that A B O systems in monkeys can only be detected by using saliva, but this is not entirely correct because agglutinations have also been obtained with red blood cells.

Among monkeys the blood groups of *Macaca sylvana* have recently been studied by Socha et al. (1981), and several simian type blood types were detected for this species. Human type blood groups were not found. Some species-specific antigens allowed the authors to point out possible taxonomic implications.

The A B O blood groups in the great apes are identical to those of humans. A, B, AB, and O have been found in chimpanzees. Genus *Pan* also has two blood groups that are unique to it (Socha, 1981). Blood group O has been demonstrated only for chimpanzees among all the apes. The *Gorilla* is most different from humans in that it has only the type B blood group.

Of the M N blood groups that are found in humans, the M antigens are quite common in many of the nonhuman primates. N antigens have been demonstrated in *Pan, Gorilla* and *Hylobates* (Landsteiner and Wiener, 1937).

Transferrins are β-globulins that are found in blood serum; they are iron-ion carriers. It appears that the transferrins of primates are rather diversified (Buettner-Janusch, 1963 a, b). More variation of the genetic polymorphism has been discovered among nonhuman

Table 10-3 A B O Blood Groups Distribution Among Primates (Modified from Erskine and Socha, 1978)

Species	O	A	B	AB	TOTAL
Alouatta spec.	-	-	52	-	52
Cebus albifrons	1	-	3	-	4
Cebus apella	-	5	-	-	5
Ateles spec.	1	10	4	-	15
Saimiri sciurea	1	3	-	-	4
Callithrix spec.	-	45	-	-	45
Papio anubis	-	5	133	56	194
Papio hamadryas	-	15	107	50	172
Papio cynocephalus	-	18	20	22	60
Papio ursinus	-	4	59	26	89
Papio papio	2	27	93	66	188
Papio (Mandrillus) leucophaeus	-	4	-	-	4
Macaca mulatta	-	-	150	-	150
Macaca radiata	-	18	12	15	45
Macaca fascicularis	1	23	19	19	62
Macaca speciosa	-	-	14	-	14
Macaca nemestrina	87	18	10	3	118
Macaca maura	1	23	2	-	26
Theropithecus gelada	18	-	-	-	18
Erythrocebus patas	-	26	-	-	26
Cercopithecus pygerythrus					
Ethiopia	-	126	1	1	128
South Africa	-	39	10	10	59
Pan troglodytes	50	483	-	-	533
Pan paniscus	-	9	-	-	9
Gorilla gorilla gorilla	-	-	23	-	23
Gorilla gorilla beringei	-	-	4	-	4
Pongo pygmaeus	-	41	14	16	71
Hylobates spec.	-	27	59	54	140
Symphalangus syndactylus	-	-	2	-	2

primates than is known in humans. For example, 24 different transferrin phenotypes have been described for *Lemur fulvus*, 4 for *Galago crassicaudatus*. 12 to 14 different molecular forms of transferrin proteins have been found in *Macaca*, and 34 phenotypes are known for this genus, but there may be many more. Other Old World monkeys have been found to exhibit genetic polymorphism of transferrins also. The same is true for the greater and lesser apes.

Haptoglobin, another important blood serum protein (α_2-globulin) has been found only in one single type among these species of primates that have been studied (haptoglobin allele HP 1-1), whereas in humans three haptoglobin types are known (haptoglobin allele HP 1-1, Hp 2-1, and Hp 2-2).

Appendix

In the 1960s Dr. Willi Hennig, a German zoologist specializing in insects, published a number of papers and books outlining a "new" approach to taxonomy and classification with widespread implications. In his partly philosophical publications Hennig reevaluated biological systematics and redefined and replaced different idioms that had long been widely accepted; these included such terms as "analogy" and "homology." Hennig's basic views make up an attractive internally self-consistent system; suddenly, it became fashionable in Europe in the mid-60s to write papers with a "Hennigian" viewpoint.

In Europe, after a few years this new taxonomy and terminology nearly vanished from zoological and primatological literature. Hennig's principal book, *Phylogenetic Systematics,* was later translated (after the usual lag time) into English in 1966, and by the early seventies, his taxonomic viewpoint had become fairly popular in the United States. The primary viewpoint of his followers, or "phylogenetic systematists," is that classification *must* reflect phylogeny. This contrasts with "evolutionary systematics," in which classification closely reflects phylogeny but also "the practical needs of discussion and communication" (Simpson, 1963:25). This type of classification recognizes the fact that rates of evolution vary and includes this notion in its classificatory schema. A noteworthy example is the classification of the hominoid primates—apes and humans. The true evolutionary relationship of these forms are

a.) Hylobates and Symphalangus

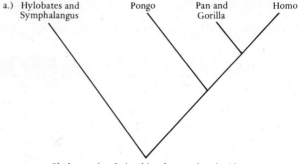

Phylogenetic relationship of extant hominoids.

b.) from Simpson (1963).

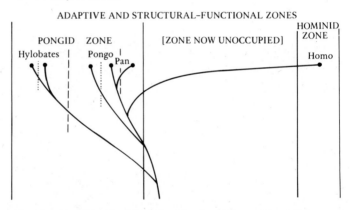

Figure A–1 Contrasting classifications by phylogenetic systematists and evolutionary systematists.

shown in the phylogeny of Figure A–1a. In contrast, Figure A–1b diagrams the point that *Homo* has undergone more evolutionary change than have *Pan/Gorilla* since their lineages diverged. Phylogenetic classification would ignore the information expressed in Figure A–1b and would tabulate the relationships something like this:

(1) Hominoidea
    2. Hylobatidae
        3. Hylobatinae
            4. *Hylobates, Symphalangus*

> 2. Pongidae
> > 3. "Pongoinae"
> > > 4. *Pongo*
> > 3. "Paninae"
> > > 4. *Pan, Gorilla*
> > > 4. *Homo*

This classification, for example, does not tell us that although *Homo* and the African great apes are more closely related, the African apes and *Pongo* are structurally and adaptively more similar to one another then both are to *Homo*. In contrast, the classification used in this book is:

> (1) Hominoidea
> > 2. Hylobatidae
> > > 3. *Hylobates, Symphalangus*
> > 2. Pongidae
> > > 3. *Pongo, Pan, Gorilla*
> > 2. Hominidae
> > > 3. *Homo*

This classification includes the information about adaptive and structural similarities of hominoids presented in Figure A-1b. Because of this flexible nature of evolutionary classification (i.e. it includes information about the animals classified beyond their phylogeny), it is preferred by the author over Hennig's phylogenetic classification and followed in this text. See also Rempe (1968) and Thenius (1970).

Bibliography

Ankel, F. 1970. *Einführung in die Primatenkunde.* Stuttgart: Gustav Fischer.

———. 1972. Vertebral morphology of fossil and extant primates: Methods of study and recent advances, in *The functional and evolutionary biology of primates.* ed. R.H. Tuttle. Chicago, New York: Aldine, Atherton.

Aldrich-Blake, F.P.G. 1968. A fertile hybrid between two *Cercopithecus* spp. in the Budongo Forest. *Folia Primatol.,* 9: 15.

Altner, G. 1968. Histologisch und vergleichend anatomische Untersuchungen zur Ontogenie und Phylogenie des Handskelettes von *Tupaia glis* (Diard 1820) und *Microcebus murinus* (J.F. Miller 1777). Ph.D. dissertation at University Giessen, Germany.

Ashton, E.H., and Oxnard, C.E. 1963. The musculature of the primate shoulder. *Trans. Zool. Soc. Lond.* 29: 553–650.

———. 1964. Locomotor patterns in primates. *Proc. Zool. Soc.* London, 142: 1–28.

Badrian, A., and Badrian, N.L. 1977. Pygmy chimpanzee. *Oryx* 13: 463.

Bauchop, T., and Martucci, R.W. 1968. Ruminant like digestion of the langur monkey. *Science* 161: 698.

Bender, M.A., and Chu, E.H.Y. 1963. The chomosomes of primates, in *Evolutionary and genetic biology of primates.* ed. J. Buettner-Janusch. vol. II. 261. New York: London: Academic Press.

Bennejeant, Ch. 1935. Les dentures temporaires des Primates. *Bull. Mém. Soc. Anthr.* Paris. 4: 11.

Bernischke, K. 1969. *Comparative mammalian cytogenetics.* New York: Springer.

Bernischke, K., and Bogart, M.H. 1976. Chromosomes of the Tan-Handed Titi (*Callicebus torquatus*), Hoffmannsegg, 1807. *Folia Primatol.* 25: 25–34.
Bernstein, I.S. 1966. Naturally occurring primate hybride. *Science.* 154: 1559.
Bernstein, R., and M. Pinto, G. Morcom, C. Bielert. 1980. A reassessment of the karotype of *Papio ursinus.* Homology between human chromosome 15 and 22 and a characteristic submetacentric baboon chromosome. *Cytogenet. Cell Gene.* 28: 55–63
Biegert, J. 1961. Volarhaut der Hände und Füsse, *Primatologia III,* 326. Basel, New York: Karger.
———. 1963. The evaluation of characteristics of the skull, hands, and feet for primate taxonomy, in *Classification and human evolution.* ed. S.L. Washburn. Chicago: Aldine Atherton.
Bierens de Haan, J.A., and Frima, M.J. 1930. Versuche über den Farbensinn der Lemuren, *Z. vergl. Physiol.* 12: 603.
Blaffer-Hrdy, S. 1977. *The langurs of Abu.* Cambridge, Mass.: Harvard Univ. Press.
Bluntschli, H. 1938. Die Sublingua und Lyssa der Lemuridenzunge. *Biomorphosis* 1: 127.
de Boer, L.E.M. 1973a. Studies on the cytogenetics of prosimians. *J. Hum. Evol.* 2: 271–278.
———. 1973b. Cytotaxonomy of the Lorisoidea (Primates: Prosimii) II. Chromosome studies in the Lorisidae and karyological relationships within the superfamily. *Genetica* 44 :330–367.
———. 1975. The somatic chromosome complement and the ideogram of *Pithecia pithecia pithecia* (Linnaeus, 1766). *Folia Primatol.* 23 :149–157.
Bogart, M.H., and Bernischke, K. 1977. Chromosome analysis of the chimpanzee (*Pan paniscus*) with a comparison to man. *Folia Primatol.* 27: 60–67.
Bolk, L. 1914. *Die Morphogenie der Primatenzähne.* Jena: Gustav Fischer.
———. 1915. Welcher Gebissreihe gehören die Molaren an? *Z. Morph. Anthr.* 17:83.
———. 1926. *Das Problem der Menschwerdung.* Jena: Gustav Fischer.
Boswell, J. 1786. *A Journal of a tour to the Hebrids.* London.
Brues, A. 1976. *People and races.* New York: Macmillan.
Buettner-Janusch, J. 1963a. *Physical anthropology: a perspective.* New York: Wiley.
———. 1963b. Hemoglobins and transferrins of baboons. *Folia Primatol.* 1: 73–87.
deCaballero, O.M.T., Ramerez, C., and Yunis, E. 1976. Genus *Cebus* Q- and G-band karyotypes and natural hybrids. *Folia Primatol.* 26: 310–321.
Carlson, A. 1922. Über die Tupaiidae und ihre Beziehungen zu den Insectivoren und den Prosimiae. *Acta Zool.* 3: 227.
Carpenter, C.R. 1934. A field study of the behavior and social relations of howling monkeys. *Comp. Psychol. Monogr.* 10:1.

———. 1940. A field study in Siam on the behavior and social relations of the gibbon (*Hylobates lar*). *Comp. Psychol. Monogr.* 16:1.

Cartmill, M. 1972. Arboreal adaptations and the origin of the order Primates, in *The functional and evolutionary biology of primates*. ed. R.H. Tuttle. Chicago, New York: Aldine Atherton.

———. 1975. Strepsirhine basicranial structures and affinities of Cheirogaleidae, in *Phylogeny of primates*. eds. W. Luckett and F. Szalay, 313, New York: Plenum.

Cartmill, M., and Milton, K. 1977. The lorisiform wrist joint and the evolution of "brachiating" adaptations in the Hominoidea. *Amer. Journ. Phys. Anthr.* 47: 249.

Casperson, T., Zech, L. and Johansson, C. 1970. Differential binding of alkylating fluorochromes in human chromosomes. *Exp. Cell Res.* 60: 315.

Chalmers, N.R. 1968. Group composition ecology and daily activities of free living mangabeys in Uganda. *Folia Primatol.* 263: 258.

Charles-Dominique, P. 1977. *Ecology and behavior of nocturnal primates: Prosimians of equatorial West Africa.* New York: Columbia Univ. Press.

Chiarelli, B. 1973. Check-list of Catarrhina: primate hybrids. *J. Human Evol.* 2: 301.

Chivers, D.J. 1974. The siamang of Malaya. *Contr. Primat.* 4: 1.

Chivers, D.J. ed. 1980. *Malayan forest primates.* New York: Plenum.

Christen, A. 1968. Haltung und Brutbiologie von *Cebuella*. *Folia Primatol.* 8: 41–49.

Chu, E.H.Y., and Bender, M.A. 1961. Chromosome cytology and evolution in primates. *Science.* 133: 1399–1405.

Chu, E.H.Y., and Bender, M.A. 1963. Cytogenetics and evolution of primates. *Ann. NY Acad. Sci.* 102: 253–266.

Coimbra-Filho, A.F., and Mittermeier, R.A. 1978. Tree-gouging, exudate eating, and the "short tusked" condition in *Callithrix* and *Cebuella*, in *The Biology and Conservation of the Callitrichidae.* ed. D.G. Kleiman. Smithsonian Institution Press, 105–118.

Collias, N., and Southwick, C. 1952. A field study of population density and social organization in howling monkeys. *Proc. amer. phil. soc.* 96: 143.

Conaway, C.H., and Sade, D.S. 1965. The seasonal spermatogenic cycle in free ranging rhesus monkeys. *Folia Primatol.* 3: 1–12.

Conroy, G.C. 1974. Primate postcranial remains from the Fayum province, Egypt, U.A.R. Ph.D. dissertation at Yale University.

Conroy, G.C., and Fleagle, J.G. 1972. Locomotor behavior in living and fossil pongids. *Nature* 237: 103.

——— and Wible, J. 1978. Middle ear morphology of *Lemur variegatus*: some implications for primate paleontology. *Folia Primatol.* 29: 81.

Cope, E.D. 1889. On the trituburcular molar in human dentition. *Journ. Morph.* 21: 7.

Crook, J.H., and Aldrich-Blake, P. 1968. Ecological and behavioral contrasts between sympatric ground dwelling primates in Ethiopia. *Folia Primatol.* 8: 192–227.

Cuvier, G. 1835. *Leçons d'anatomie comparée*. vol. 1–5. Paris.
Cuvier, G., et Geoffroy, E. 1795. *Mém. Class. Mamm.* Paris.
Dahlberg, A.A. 1971. *Dental morphology and evolution*. Chicago: Univ. of Chicago Press.
Doran, A.H.G. 1876a. On the comparative anatomy of the auditory ossicles of the mammalia. *Proc. Roy. Soc. London* 25: 101.
———. 1876b. Morphology of the mammalian ossicula auditus. *Transact. Linn. Soc. London, Zool.* 1: 371.
Dresser, M.E., and Hamilton, A.E. 1979. Chromosomes of Lemuriformes. V. A comparison of the karyotypes of *Cheirogaleus medius* and *Lemur fulvus fulvus*. *Cytogenet. Cell Genet.* 24: 160–167.
Dukelow, W.R. 1978. The nonhuman primate as a reproductive model for man. *Symp. Species Selection in Biomedical Primatology, A.S.P.* Atlanta.
Dutrillaux, B., Viegas-Pequignot, E., Dubos, C., and Masse, R. 1978a. Complete or almost complete analogy of chromosome banding between the baboon (*Papio papio*) and man. *Hum. Genet.* 43:37–46.
Dutrillaux, B., Viegas-Pequignot, E., Couturier, J., and Chauvier, G. 1978b. Identity of euchromatic bands from man to cercopithecidae. *Hum. Genet.* 45: 283–296.
Dutrillaux, B., Biemont, M.C., Viegas-Pequignot, E., and Laurest, C. 1979. Comparison of the karyotypes of four Cercopithecoidea: *Papio papio, P. anubis, Macaca mulatta, M. fascicularis. Cytogenet. Cell Genet.* 23: 77–83.
Dutrillaux, B., Couturier, J., and Fosse, A.M. 1980. The use of high resolution banding in comparative cytogenetic. Comparison between man and *Lagothrix lagotricha* (Cebidae). *Cytogenetics.* 27: 45–51.
Edinger, T. 1961. Anthropocentric misconceptions in palaeoneurology. *Proc. Rudolf Virchow Med. Soc.* New York. 19: 56.
Egozcue, J. 1969. Primates, in *Comparative mammalian cytogenetics*. ed. K. Bernischke. New York: Springer.
Egozcue, J., Perkins, M., and Hagemenas, F. 1968. Chromosomal evolution in marmosets, tamarins, and pinches. *Folia Primatol.* 9: 81–94.
Ellefson, J.O. 1968. Territorial behavior in the common white handed gibbon, *Hylobates lar* Linn., in *Primates*. ed. H. Jay. New York: Holt, Rinehart and Winston.
Epple, G. 1967. Vergleichende Untersuchungen über Sexual- und Socialverhalten der Krallenaffen (Hapalidae). *Folia Primatol.* 7: 37.
Epple, G., und Lorenz, R. 1967. Vorkommen, Morphologie und Funktion der Sternaldrüse bei den Platyrrhini. *Folia Primatol.* 7: 98.
Erikson, G.E. 1963. Brachiation in New World monkeys and in anthropoid apes. *Symp. Zool. Soc. London.* 10: 135.
Erskine, A.G. and Socha, W.W. 1978. *The principles and practice of blood grouping*. St. Louis: The C.V. Mosby Company.
Every, G., 1970. Sharpness of teeth in man and other primates. Postilla, 143, 1.
Fiedler, W. 1956. Übersicht über das System der Primaten. *Primatologia. I*: 1. Basel, New York: Karger.

Fleagle, J.G. 1977a. Locomotor behavior and muscular anatomy of sympatric malaysian leaf-monkeys (*Presbytis obscura* and *Presbytis melalophos*). *Amer. Journ. Phys. Anthr.* 46: 297.
———. 1977b. Locomotor behavior and skeletal anatomy of sympatric malaysian leaf-monkeys (*Presbytis obscura* and *Presbytis melalophos*). *Yearbook Phys. Anthr.* 20: 440.
Fleagle, J.G. and Mittermeier, R.A. 1980. Locomotor behavior, body size, and comparative ecology of seven Surinam monkeys. *Amer. Journ. Phys. Anthr.* 52: 301–314.
Fooden, J. 1969. Taxonomy and evolution of the monkeys of Celebes (Primates, Cercopithecidae). *Bibl. Primat.* 10. Basel, New York: Karger.
Ford, C.E. and Hamerton, J.L. 1956. The chromosomes of man. *Nature* 178:1,020–1,023.
Ford, C.E. and Evans, E.P. 1977. Cytogenetic observations on chimerism in heterosexual twin cattle and marmosets. *Journ. Reprod. Fert.* 49: 25–31.
Fossey, D. 1976. *The behavior of the mountain gorilla.* Ph.D. dissertation at Cambridge University.
Friant, M. 1935. Description et interpretation de la dentition d'un jeune Indris. *C.R. Assoc. Anat.* 30: 205–213.
Galdikas-Brindamour, B. 1975. Orangutans, Indionesia's "people of the forest." *Nat. Geogr. Soc. Journ.* 148: 444.
Garcia, M., Miro, R., and Egozcue, J. 1978. Banding patterns of the chromosomes of *Nycticebus coucang* (Boddaert, 1785). *Folia Primatol.* 29: 103–106.
Gartlan, J.S. 1975. Adaptive aspects of social structure in *Erythrocebus patas*, in *Proc. Symp. 5th Congress Intern. Primat. Soc.* Nagoya, Japan. 161.
Garver, J.J., Estop, A. Pearson, P.L., Dijksman, T.M., Wijnen, L.M.M., and Meera Khan, P. 1977. Comparative gene mapping in the Pongidae and Cercopithecoidea, in *Chromosomes Today.* 6: 191–199. ed. A. dela Chapelle.
Gasden, J.R., Mitchell, A.R., and Seuanez, H.N. 1977. Localisation of sequences complementary to human satellite DNAs in man and the hominoid apes, in *ICN-UCLA Symp. Molec. and Cell Biol.* VII: 75–85. eds. R.S. Sparks, D.E. Comings, and C.F. Fox.
Geoffroy, E. 1812. *Ann. Mus. Hist. Nat.* Paris XIX: 157.
Gérard, P. 1932. Études sur l'ovogénèse et l'ontogénèse chez les Lémuriens du genre *Galago. Arch. Biol.* Paris. 43: 93.
Gingerich, P.D. 1974. *Cranial anatomy and evolution of early tertiary Plesiadapidae (Mammalia, Primates).* Ph.D. dissertation at Yale University.
Gingerich, P.D. 1977. Homologies of the anterior teeth in Indriidae and a functional basis for dental reduction in primates. *Amer. Journ. Phys. Anthr.* 47: 387.
Glaser, D. 1970. Über die Ossification der Extremitäten bei neugeborenen Primaten (Mammalia). *Z. Morph. Tiere* 68: 127.

———. 1972a. Vergleichende Untersuchungen über den Geschmackssinn der Primaten. *Folia Primatol.* 17: 267.

———. 1972b. Zur Problematik der Untersuchungen mit dem Geschmacksstoff Phenylthiocarbamid (PTC). *Z. Morph. Anthr.* 64: 197.

Goodall, J. 1965. Chimpanzee of the Gombe stream reserve, in *Primate behavior.* ed. I. De Vore. New York: Holt, Rinehart and Winston.

———. 1968. The behavior of free living chimpanzees in the Gombe stream reserve. *Anim. Beh. Monogr.* 1: 161.

———. 1979. Life and death at Gombe. *Nat. Geogr. Soc. Journ.* 155: 593.

Goodman, M. 1961. The role of immunochemical differences in the phyletic development of human behavior. *Human biology* 33: 131.

———. 1962a. Evolution of the immunologic species specifity of human serum proteins. *Human biology* 34: 104.

———. 1962b. Immunochemistry of the primates and primate evolution. *Science* 102: 219.

Gould, S.J. 1975. Allometry in primates, with emphasis on scaling and the evolution of the brain, in *Approaches to primate paleobiology.* ed. F. Szalay, Contr. primat. 5: 244. Basel, New York: Karger.

Grand, T. 1968. Functional anatomy of the upper limb, in *Biology of the howler monkey (Alouatta caraya).* ed. M.R. Malinow, Biblio. primat. 7: 104. Basel, New York: Karger.

Gregory, W.K. 1916. Studies on the evolution of the primates. *Amer. Mus. Nat. Hist.* 35: 239.

———. 1922. *The origin and evolution of the human dentition.* Baltimore: Williams and Wilkins.

———. 1934. *Man's place among anthropoids.* Clarendon Press, Oxford.

——— and Hellman, M. 1926. The dentition of *Dryopithecus* and the origin of man. *Anthr. Papers, Amer. Mus. Nat. Hist.* 28: 123.

Gropp, A. 1970. Cytologische Mechanismen der Evolution der Säugerkaryotypen. *Anthr. Anz.* 32: 220.

Grosser, O. 1909. *Vergleichende Anatomie und Entwicklungsgeschichte der Eihäute und der Plazenta.* Leipzig: Wien.

Groves, C.P. 1967. Ecology and taxonomy of the gorilla. *Nature.* London. 213: 890.

Haines, R.W. 1950. The interorbital septum in mammals. *Journ. Linn. Soc.* London. 41: 585.

Halaczek, B. 1972. *Die Langknochen der Hinterextremität bei simischen Primaten, eine vergleichend morphologische Untersuchung.* Zürich: Juris Druck und Verlag.

Haller, A. von. 1762. *Elementa physiologicae corpus humani.* Lausanne.

Hamilton, A.E., and Buettner-Janusch, J. 1977. Chromosomes of Lemuriformes III. The genus *Lemur:* karyotypes of species, subspecies and hybrids. *Ann. NY Acad. Sci.* 293: 125–159.

Harlow, H.F., and Harlow, M.K. 1962. Social deprivation in monkeys. *Scient. Amer.* 207:136.

Heffner, H.E., Ravizza, R.J., and Masterton, B. 1969a. Hearing in primitive mammals, III, tree shrew *(Tupaia glis). Journ. audit. Res.* 9: 12.

———. 1969b. Hearing in primitive mammals IV, bush baby (*Galago senegalensis*). *Journ. audit. Res.* 9: 19.

Hellekant, G., Glaser, D., Brouwer, J.N., and van der Vel, H. 1976. Gustatory effects of Miraculin, Monellin and Thaumatin in the *Saguinas midas* tamarin monkey, studied with electro-physiological and behavioral techniques. *Acta physiol. scand.* 97: 241.

Hemprich, W. 1820. *Grundriss der Naturgeschichte für höhere Lehranstalten.* August Rücker, Berlin und Friedrich Völke, Wien.

Hennig, W. 1966. *Phylogenetic Systematics.* Chicago: University of Illinois Press.

Herwerden, van, M. 1906. Die puerperalen Vorgänge in der Mucosa von *Tupaia javanica. Anat. Hefte* 32: 155.

Hershkovitz, P. 1971. Basic crown patterns and cusp homologies in mammalian teeth, in *Dental morphology and evolution.* ed. A. Dahlberg, 95. Chicago: The Univ. of Chicago Press.

Hesse, R. 1928. Die Ohrmuscheln des Elephanten als Wärmeregulatoren. *Zeitschr. wiss. Zool.* 132: 314.

Hill, W.C.O. 1953. *Primates. Comparative anatomy and taxonomy.* vol. I, Strepsirhini. Edinburgh: Edinburgh Univ. Press.

Hofer, H.O. 1958. Über das Bewegungsspiel der Klammeraffen. *Natur und Volk* 90: 241–251.

———. 1965. Die morphologische Analyse des Schädels des Menschen, in *Menschliche Abstammungslehre, Fortschritte der Anthropogenie 1863–1964.* ed. G. Heberer, 145. Stuttgart: Gustav Fischer.

———. 1969. On the organon sublinguale in *Callicebus* (Primates, Platyrrhini). *Folia Primatol.* 11: 268.

———. 1971. Comparative studies on the sublingual organ in primates. The sublingua in *Perodicticus potto,* (Primates, Prosimiae, Lorisiformes), in *Proc. 3rd Congr. Intern. Primat. Soc. Zürich 1970.* vol. I, 198. eds. J. Biegert and W. Leutenegger, Basel, New York: Karger.

———. 1972. A comparative study on the oro-nasal region of the external face of the Gorilla as a contribution to cranio-facial biology of the primates. *Folia Primatol.* 18: 416.

——— and Altner, G. 1972. *Die Sonderstellung des Menschen.* Stuttgart: Gustav Fischer.

Holloway, R.L. 1966. Cranial capacity, neural reorganization and hominid evolution. *Amer. Anthr.* 68: 103.

———. 1972. Australopithecine endocasts, brain evolution, in the Hominoidea and a model of hominid evolution, in *The functional and evolutionary biology of primates.* ed. R.H. Tuttle, 185. Chicago, New York: Aldine Atherton.

Horn, D.A. 1976. A preliminary report on the ecology and behavior of the bonobo chimpanzee (*Pan paniscus* Schwartz, 1929) and a reconsideration of the evolution of the chimpanzees. Ph.D. dissertation at Yale University.

Huxley, T.H. 1863. *Evidence as to man's place in nature.* London: William and Norgate.

Hylander, W.L. 1975. Incisor size and diet in anthropoids with special reference to Cercopithecidae. *Science* 189: 1095–1098.
James, W.W. 1960. *The jaws and teeth of primates.* London: Pitman Medical Publishing Company.
Jenkins, F.A. Jr. 1970. Anatomy and function of expanded ribs in certain edentates and primates. *Journ. Mammal.* 51: 288.
———. 1974. *Primate Locomotion.* Academic Press, New York, London.
———. 1981. Wrist rotation in primates: A critical adaptation for brachiators. *Symp. Zool. Soc.* London, 48: 429–451.
Jenkins, F.A., Dombrowski, P.J., and Gordon, E.P. 1978. Analysis of the shoulder in brachiating spider monkeys. *Amer. Journ. Phys. Anthr.* 48: 65.
Jolly, A. 1972. *The evolution of primate behavior.* New York: Macmillan.
Jolly, C.J. 1970. The seed-eaters: a new model of hominid differentiation based on a baboon analogy. *Man* 5: 5.
Jones, M.L. 1968. Longevity of primates in captivity. *Intern. Zoo Yearbook,* 8, Publ. Zool. Soc. London.
Kay, R.F. 1973. Mastication, molar tooth structure and diet in primates. Ph.D. dissertation at Yale University.
———. 1975. The functional adaptations of primate molar teeth. *Am. J. Phys. Anthrop.* 43: 195–216.
———. 1978. Molar structure and diet in extant Cercopithecidae, in *Development, Function, and Evolution of Teeth.* eds. P.M. Butler and K.A. Joysey, New York: Academic Press.
——— and Hiiemae, K.M. 1974. Mastication in *Galago crassicaudatus,* a cinefluoro-graphic and occlusal study, in *Prosimian biology.* eds. R.D. Martin, G.A. Doyle, and A.C. Walker, 501. London: Duckworth.
Keith, A. 1902. The extent to which the posterior segments of the body have been transmuted and suppressed in the evolution of man and allied primates. *Journ. Anat.* London, 37: 18–40.
King, M.C., and Wilson, A.C. 1975. Evolution at two levels in humans and chimpanzees. *Science* 188: 107.
Klinger, H.P. 1963. The somatic chromosomes of some primates (*Tupaia glis, Nycticebus coucang, Tarsius bancanus, Cercocebus aterrimus, Symphalangus syndactylus*) *Cytogenet.* 2: 140–151.
Kohne, D. 1970. Evolution of higher organism DNA. *Quart. Rev. Biophysics* 3: 327.
Kollman, M. 1925. Études sur les lémuriens: la fosse orbitotemporale et l'os planum. *Mém. Soc. Linn. Normandie (Zool.)* 1: 3.
Kuhn, H.J. 1964. Zur Kenntnis von Bau und Funktion des Magens der Schlankaffen (Colobinae). *Folia Primatol.* 2: 193.
———. 1967. Zur Systematik der Cercopithecidae, Neue Ergebnisse der Primatologie. *First Congr. Intern. Primat. Soc. Frankfurt, Main 1966* 25. Stuttgart: Gustav Fischer.
Landsteiner, K., and Wiener, A.S. 1937. M-reagents in apes and monkeys. *Journ. Immunol.* 33: 19–22.

Lasinsky, W. 1960. Äusseres Ohr. *Primatologia II* 41. Basel, New York: Karger.
Lau, Y.-F., and Arrighi, F.E. 1976. Studies on the squirrel monkey, *Saimiri sciureus*, genome I. Cytological characterizations of chromosomal heterozygosity. *Cytogenet. Cell Genet.* 17: 51–60.
Leche, W. 1896. *Untersuchungen über das Zahnsystem lebender und fossiler Halbaffen. Festschrift für Gegenbauer.* Leipzig.
Leutenegger, W. 1973. Encephalization in australopithecines: a new estimate. *Folia Primatol.* 19: 9.
——. 1974. Functional aspects of pelvic morphology in simian primates. *Journ. Human Evol.* 3:207.
Lieberman, P.H. 1975. *On the origin of language.* New York: Macmillan.
Lorenz, R. 1971. The functional interpretation of the thumb in the Hylobatidae, in *Proc. third Congr. Intern. Primat. Soc. Zürich 1970.* vol. I, 130 eds. J. Biegert, and W. Leutenegger, New York: Karger, Basel.
Lovejoy, C.O. 1972. Primate phylogeny and immunological distances. *Science* 176: 803.
Luckett, W.P. 1968. Morphogenesis of the placenta and fetal membranes of the tree shrews (Family Tupaiidae). *Amer. Journ. Anat.* 123: 383.
——. ed. 1980. *Comparative Biology and Evolutionary Relationships of Tree Shrews.* New York: Plenum.
Lyon, M.W. 1913. Tree shrews. An account of the mammalian family Tupaiidae. *Proc. U.S. Natl. Museum* 45: 1.
Ma, N.F.F., Johnes, T.C., Thorington, R.W., and Cooper, R.W. 1974. Chromosome banding pattern in squirrel monkeys (*Saimiri sciureus*). *J. of med. primatol.* 3: 120.
Mac Phee, R. 1977. Ontogeny of the ecto-tympanic-petrosal plate relationship in strepsirhine prosimians. *Folia Primatol.* 27: 245.
Malbrant, R., and Maclatchy, A. 1949. *Faune de l'équateur Africain Francais, Tome II, Mammifères.* Paris: Paul Lechevalier.
Martin, R.D. 1968. Towards a new definition of primates. *Man.* 3: 377.
——. 1972. Adaptive radiation and behavior of the malagasy lemurs. *Phil. Trans. Roy. Soc. London,* 8: 264, 295.
Masali, M. 1968. The earbones and the vertebral column as indicators of taxonomic and postural distinction among Old World primates, in *Taxonomy and phylogeny of Old World primates.* ed. B. Chiarelli, 69. Torino: Rosenberg and Sellier.
McArdle, J.E. 1981. *Functional morphology of the hip and thigh of the Lorisiformes.* Contributions in Primatology, 17. Karger, Basel.
Mendel, F. 1976. Postural and locomotor behavior in *Alouatta palliata* on various substrates. *Folia Primatol.* 26: 36–56.
Miller, D.A. 1977. Evolution of primate chromosomes *Science.* 198:1116.
Miller, C.K., Miller, D.A., Miller, O.J., Tantravahi, R., and Reese, R.T. 1977. Banded chromosomes of the owl monkey, *Aotus trivirgatus. Cytogenet. Cell Genet.* 19: 215–226.
Mittermeier, R.A. 1978. Locomotion and posture in *Ateles geoffroyi* and *Ateles paniscus. Folia Primatol.* 30: 161–193.

Mittermeier, R.A., and Fleagle, J.G. 1976. The locomotor and postural repertoire of *Ateles* geoffroyi and *Colobus guereza* and a reevaluation of the locomotor category semibrachiation. *Amer. Journ. Phys. Anthr.* 45: 235.

Mivart, S.G. 1873. On *Lepilemur* and *Cheirogaleus* and the zoological rank of the Lemuroidea. *Proc. zool. Soc.* London. 484: 510.

Napier, J.R. 1961. Prehensibility and opposability in the hands of primates. *Symp. Zool. Soc.* London. 5: 115.

—— and Napier, P.H. 1967. *A handbook of living primates.* London, New York: Academic Press.

—— and Walker, A.C. 1967. Vertical clinging and leaping. A newly recognized category of locomotor behavior of primates. *Folia Primatol.* 7: 204.

Nishida, T. 1968. the social group of wild chimpanzees in the Mahali Mountains. *Primates* 9: 167.

Noback, Ch. R., and Moskowitz, N. 1963. The primate nervous system: functional and structural aspects of phylogeny, in *Evolutionary and genetic biology of primates.* ed. J. Buettner-Janusch. vol. I. 131. London, New York: Academic Press.

Ohno, S. 1969. The mammalian genome in evolution and conservation of the original x-linkage group, in *Comparative mammalian Cytogenetics.* ed. K. Bernischke, 18. New York: Springer.

Osborn, H.F. 1907. The Fayum expedition of the American Museum *Science* 25: 513.

Owen, R. 1866. On the Aye-Aye (Chiromys Cuvier). *Trans. Zool. Soc. Lond.* 5: 33–101.

Oxnard, C.E. 1972. African fossil foot bones. *Amer. Journ. Phys. Anthr.* 37: 3.

Paris Conference 1975. Birth defects: original article series, Suppl. 11: 9, (The National Foundation, NY), 1971.

Patterson, B. 1956. Early cretaceous mammals and the evolution of mammalian molar teeth. *Fieldiana Geol.* 13: 105.

Pearson, P., Garver, J., Estop, A., and Dijksman, T. 1977. The chromosomes of non-human primates, in *ICN-UCLA Symp. on Molec. and Cell Biol.* VII 415–425, eds. R.S. Sparks, D.E. Comings, and C.F. Fox.

Peters, W., 1866. Über die Säugetiergattung *Chiromys*, Abh. Acad. der Wiss. Berlin.

Petter, J., 1962. Ecological and behavioral studies of madagascan lemurs in the field, Ann. N.Y. Acad. Sci., 102, 267.

Petter, J.-J., Schilling, A, and Pariente, 1971. Observations éco-ethologiques sur deux lemuriens malgaches nocturnes: *Phaner furcifer* et *Microcebus coquereli.* Terre et Vie 25, 287.

Petter, J.J., Albignac, R., and Rumpler, Y. 1977. Faune de Madagascar, Mammifères, Lémuriens (Primates, Prosimiens), Faune de Madagascar 44, Orstom. *CNRS.* Paris.

Petter-Rousseaux, A. 1968. cycles génitaux saisonniers des lémuriens malgaches, in *Cycles genitaux saisonniers de mammifères sauvages.* ed. R. Canivenc, 11. Paris.

Pilbeam, D.R. 1972. *The ascent of man, an introduction to human evolution.* New York: Macmillan.
Pocock,R.I. 1918. On the external characters of lemurs and of *Tarsius*. *Proc. Zool. Soc. London* 19: 53.
Pollock, J. 1976. The social behavior and ecology of *Indri indri.* Ph.D. Dissertation at University College, London.
———. 1977. The ecology and sociology of feeding in *Indri indri*, in *Primate Ecology*, Ed. Clutton-Brock. New York: Academic Press.
Radinsky, L. 1972. Endocasts and studies of primate brain evolution, in *The functional and evolutionary biology of primates.* ed. R.H. Tuttle, 175. Chicago, New York: Aldine Atherton.
Remane, A. 1921. Beiträge zur Morphologie des Anthropoiden Gebisses. *Wiegmann's Archiv für Naturgeschichte* Abt. A, 87.
———. 1956. Paläontologie und Evolution der Primaten. *Primatologia* I: 267. Basel, New York: Karger.
———. 1960. Zähne und Gebiss der Primaten. *Primatologia* III: 637. New York: Karger, Basel.
———. 1967. Phylogenetische Methoden ausserhalb der morphologischen Verwandtschaftsforschung (Parasitologie, Biochemie). *Zool. Anz.* 179: 80.
Rempe, U. 1968. Die Rekonstruktion von Stammbäumen auf Grund von Immunpräzipitinen dargestellt am Beispiel der Hominoidea. *Z. Morph. Anthr.* 60: 1.
Reynolds, V., and Reynolds, F. 1965. Chimpanzees in the Budongo Forest, in *Primate behavior.* ed. I. De Vore, 368. New York: Holt, Rinehart and Winston.
Richard, A. 1974. Intra-specific variation in the social organization and ecology of *Propithecus verreauxi. Folia Primatol.* 22: 178.
Riopelle, A.J. 1963. Growth and behavioral changes in chimpanzees. *Z. Morph. Anthr.* 53: 1.
Rodman, P.S. 1973. Population composition and adaptive organization among oran-utans of the Kutai reserve, in *Comparative ecology and behavior of primates.* eds. R.P. Michael, and J.H. Crook. London, New York: Academic Press.
Rohen, J.W., and Castenholz, A. 1967. Über die Zentralisation der Retina bei Primaten. *Folia Primatol.* 5: 92.
Rose, M.D. 1973. Quadrupedalism in primates. *Primates* 14: 337–357.
———. 1974. Postural adaptations in New and Old World monkeys, in *Primate Locomotion*, ed. F.A. Jenkins Jr., pp. 201–222, Academic Press, New York.
Ruch, T.C. 1941. *Primatologica, part I: anatomy, embryology and quantitative morphology.* Springfield, Illinois.
Rumpler, Y. 1974. Cytogenetics and classification of lemurs, in *Prosimian biology.* eds. R.D. Martin, G.A. Doyle, and A.C. Walker, 865. London: Duckworth.
———. 1975. The significance of chromosomal studies in the systematics of the Malagasy lemurs. In: Lemur Biology, 25–40.
Rumpler, Y., and Albignac, R., 1973. Cytogenetic study of the endemic

Malagasy Lemur: *Hapalemur*, I. Geoffroy, 1851. J. Hum. Evol., 2, 267–270.

Rumpler, Y., and Dutrillaux, B., 1976. Chromosomal evolution in Malagasy lemurs. *Cytogenet. Cell Genet.*, 17, 268–281.

Rumpler, Y., and Dutrillaux, B., 1979. Chromosomal evolution in Malagasy lemurs IV. Chromosome banding studies in the genuses *Phaner, Varecia, Lemur, Microcebus*, and *Cheirogaleus*. Cytogenet. Cell Genet., 24, 224–232.

Saban, R. 1963. Contribution à létudes des l'os temporal des primates. *Mém. Mus. nat. d'hist. nat. Paris* (Zool) 29: 1.

Sabater Pi, J. 1973. Contribution to the ecology of *Colobus polykomos satanas* (Waterhouse 1838) of Rio Muni, Republic of Equatorial Guinea. *Folia primatol.* 19: 193.

Sanides, F. 1970. Functional architecture of motor and sensory cortices in primates, in *Advances in Primatology I, the primate brain.* 137 New York: Appleton Century Crofts.

Sarich, V.M., and Wilson, A.C. 1968. Immunological time scale for hominid evolution. *Science* 158: 1200.

Schaller, G.B. 1963. *The mountain gorilla, ecology and behavior.* Chicago: Univ. of Chicago Press.

———. 1965. The behavior of the mountain gorilla, in *Primate behavior.* ed. I. De Vore, 324. New York: Holt, Rinehart and Winston.

Schneider, R. 1958. Zunge und weicher Gaumen. *Primatologia* III: 61. Basel, New York: Karger.

Schultz, A.H. 1938. The relative weight of testes in primates. *Anat. Rec.* 72: 387.

———. 1940. The size of the orbit and the eye in primates. *Amer. Journ. Phys, Anthr.* 26: 389.

———. 1948. The relation in size between premaxilla, diastema and canine. *Amer. Journ. Phys. Anthr.* 6: 163.

———. 1958. Palatine ridges. *Primatologia* III: 127, Basel, New York: Karger.

———. 1961. *Vertebral column and thorax*, in *Primatologia*, IV eds. H. Hofer, A.H. Schultz, and D. Starck. Karger, Basel.

———. 1965. Die rezenten Hominoidea, in *Menschliche Abstammungslehre.* ed. G. Heberer. Stuttgart: Gustav Fischer.

———. 1969. *The life of primates.* London: Weidenfels and Nicolson.

Schwartz, J.H. 1974. Observations on the dentitions of Indriidae. *Amer. Journ. Phys. Anthr.* 41: 107–114.

———. 1978. If *Tarsius* is not a prosimian, is it a haplorhine?, in *Recent advances in primatology*, 3, *Evolution*, eds. D.J. Chivers and K.A. Joysey, London: Academic Press, pp. 195–202.

Shine, S.S., and Kay, R.F. 1978. An analysis of chewed food particle size and its relationship to molar structure in the primates *Cheirogaleus medius* and *Galago senegalensis*, and the insectivoran *Tupaia glis*. *Amer. Journ. Phys. Anthr.* 26: 241.

Simons, E.L. 1967. Fossil primates and the evolution of some primate locomotor systems. *Amer. Journ. Phys. Anthr.* 26: 241.

Simpson, G.G. 1936. Studies on the earliest mammalian dentitions. *Dental Cosmos* 78: 791, 940.
———. 1945. The principle of classification and a classification of mammals. *Bull. Amer. Mus. Nat. Hist.* 85: 1.
———. 1963. The meaning of taxonomic statements, in *Classification and Human Evolution.* ed. S.L. Washburn. 1–31.
Socha, W.W. 1980. Blood groups of apes and monkeys. Current states and practical applications. *Lab. Animal Sci.* 30: 698–702.
———. 1981. Blood groups as genetic markers in chimpanzees: their importance for the National Chimpanzee Breeding Program. *Amer. Journ. Primat.* 1: 3–13.
Socha, W.W., Merz, E. and Moor-Jankowski, J. 1981. Blood groups of Barbary Apes (*Macaca sylvanus*). *Folia Primatol.* 36: 212–225.
Socha, W.W., and Moor-Jankowski, J. 1980. Chimpanzee R - C - E - F blood group system: a counterpart of the human Rh Hr blood groups. *Folia Primatol.* 33 172–188.
Southwick, G.H., Beg, M.A., and Siddiqi, M.R. 1965. Rhesus monkeys in north India, in *Primate behavior.* ed. I. De Vore. New York: Holt, Rinehart and Winston.
Spatz, W.B. 1964. Beitrag zur Kenntnis der Ontogenese des Craniums von *Tupaia glis* (Diard 1820). *Morph. Jahrb.* 106: 321.
———. 1968. Die Bedeutung der Augen für die sagittale Gestaltung des Schädels von *Tarsius* (Prosimia, Tarsiiformes). *Folia Primatol.* 3: 153.
Sprankel, H. 1965. Untersuchungen an *Tarsius.* I. Morphologie des Schwanzes nebst ethologischen Bemerkungen. *Folia Primatol.* 3: 153.
Spreng, H. 1938. Zur Ontogenie des Indrisinengebisses. Ph.D. dissertation at Universität Bern, Switzerland.
Starck, D. 1953. Morphologische Untersuchungen am Kopf der Säugetiere, besonders der Prosimier, ein Beitrag zum Problem des Formwandels des Säugetierschädels. *Z. Wiss. Zool.* 157: 169.
———. 1956. Primitiventwicklung und Plazentation der Primaten. *Primatologia* I: 732, Basel, New York: Karger.
———. 1960. Bauchraum und Topographie der Bauchorgane. *Primatologia* III: 446, Basel, New York: Karger.
———. 1962a. Das Cranium von *Propithecus* spec. (Prosimiae, Lemuriformes, Indriidae). *Biblio. primat.* 1. Basel, New York: Karger.
———. 1962b. Der heutige Stand des Fetalisationsproblems. *Z. Tierzüchtung und Züchtungsbiologie* 77: 1.
———. 1965. Die Neencephalisation, in *Menschliche Abstammungslehre, Fortschritte der Anthropogenie 1863–1964.* ed. G. Heberer, 103. Stuttgart: Gustav Fischer.
———. 1974. Die Stellung der Hominiden im Rahmen der Säugetiere, in Die Evolution der Organismen, ed. G. Heberer, Gustav Fischer, Stuttgart.
Stehlin, H.G. 1916. Die Säugetire des schweizerischen Eozäns. 2. Hälfte: Schlussbetrachtungen zu den Primaten. *Abh. schweiz. Pal. Ges.* 41: 1297–1552.

Steiner, H. 1951. Die embryonale Hand- und Fuss-Entwicklung von *Tupaia*. *Verhdl. schweiz. naturf. Ges.* 131: 153.

Stephan, H. 1967. Zur Entwicklungshöhe der Insectivoren nach Merkmalen des Gehirns und die Definition der "Basalen Insectivora."*Zool. Anz.* 179: 177.

———. 1972. Comparative anatomy of primate brains, in *The functional and evolutionary biology of primates* ed. R.H. Tuttle, 155. Chicago, New York: Aldine Atherton.

——— and Bauchot, R. 1965. Hirn-Köpergewichts Beziehungen bei den Halbaffen (Prosimii). *Acta Zool.* 46: 209.

Stern, J.T., and Oxnard, C.E. 1973. Primate locomotion: some links with evolution and morphology. *Primatologia* IV: 11, Basel, New York: Karger.

Stern, J.T. and Susman, R.L. 1981. Electromyography of the gluteal muscles in *Hylobates, Pongo* and *Pan:* Implications for the evolution of hominid bipedality. *Amer. Journ. Phys. Anthr.* 55: 153–166.

Sugiyama, Y. 1973. The social structure of wild chimpanzees, a review of field studies, in *Comparative ecology and behavior of primates.* eds. R.P. Michael, and J.H. Crook, 375. London: Academic Press.

Suzuki, A. 1969. An ecological study of chimpanzees in a savanna woodland. *Primates* 10: 103.

Swindler, D.R. 1976. *Dentition of living primates.* London: Academic Press.

Swindler, D.R., and Wood, C.D. 1973. *An atlas of primate gross anatomy.* Seattle and London: University of Washington Press.

Tantravahi, R., Dev, V.G., Firschein, I.L., Miller, D.A., and Miller, O.J. 1975. Karyotype of the gibbons *Hylobates lar* and *H. moloch.* Inversion in chromosome 7. *Cytogenet. Cell Genet.* 15: 92–102.

Tattersall, I. 1972. The functional significance of airorhynchy in *Megaladapis. Folia Primatol.* 18: 20.

———. 1977. Ecology and behavior in *Lemur fulvus mayottensis* (Primates, Lemuriformes). *Anthr. Pap. Amer. Mus. Nat. Hist.*54: 421.

———. 1982. *The Primates of Madagascar.* Columbia Univ. Press, New York.

Tattersall, I. and Schwartz, J. 1974. Craniodental morphology and the systematic of the Malagassy lemurs (Primates, Prosimii). *Anthro. Pap. Amer. Mus. Nat. Hist.* 52: 339.

Thenius, E. 1970. Moderne Methoden der Verwandtschaftsforschung. *Umschau* 22: 695.

Tigges, J. 1963. Untersuchungen über den Farbensinn von *Tupaia glis* (Diard 1820). *Z. Anthr. Morph.* 53: 109.

———. 1964. On visual learning capacity, retention and memory in *Tupaia glis* (Diard 1820). *Folia primatol.* 2: 232.

Tjio, J.H., and Levan, A. 1956. The chromosome number of man. *Hereditas* 42: 1.

Tuttle, R.H. 1965. The anatomy of the chimpanzee hand, with comments on hominid evolution. University Microfilms, Ann Arbor, Michigan.

———. 1969. Knuckle walking and the problem of human origins. *Science* 166: 953.

———. 1975. Knuckle walking and knuckle walkers: a commentary on some recent perspectives in hominoid evolution, in *Primate functional morphology and evolution*. ed. R.H. Tuttle. The Hague: Mouton.
Vandebroek, G. 1961. The comparative anatomy of the teeth of lower and non-specialized mammals. *Paleis der Academien*. Brussel.
Van Valen, L. 1965. Tree shrews, primates and fossils. *Evolution* 19: 137.
Vogel, C. 1962. Untersuchungen an *Colobus* Schädeln aus Liberia, under besonderer Berücksichtigung der Crista sagittalis. *Z. Morph. Anthr.* 52: 306.
———. 1868. The phylogenetical evolution of some characteristics and some morphological trends in the evolution of the skull of catarrhine primates, in *Taxonomy and phylogeny of the Old World* primates with reference to the origin of man. ed. B. Chiarelli. 21. Torino: Rosenberg and Sellier.
Vogt, G., und Vogt, O. 1907. Zur Kenntnis der elektrisch erregbaren Hirnrinden Gebiete bei den Säugetieren. *Journ. Psych. Neurol.* Leipzig. 8: 277.
Walker, A.C. 1969. The locomotion of lorises, with special reference to the potto. *E. Afr. Wildlife Journ.* 7: 1.
———. 1970. Nuchal adaptations in *Perodicticus potto.Primates* 11: 135.
——— and Rose, M.D. 1968. Fossil hominoid vertebra from the Miocene of Uganda. *Nature* 217: 980.
Watson, J.O., and Crick, F,H,C. 1953. The structure of DNA. *Cold Spring Harbor Symp. Quant. Biol.* 18: 123.
Wen, I.C. 1930. Ontogeny and Phylogeny of the nasal cartilages in primates. *Contr. Embryol.* Carnegie Inst. Washington, Publ. 414: 109.
Werner, C.F. 1960. Mittel- und Innenohr. *Primatologia* II: 1. Basel, New York: Karger.
Wettstein, E.B. 1963. Geschlechtsunterschiede und Altersveränderungen bei *Callithrix jacchus.Morph. Jb.* 104: 185.
Wiedersheim, R. 1902. *Anatomie der Wirbelthiere*. Jena: Gustav Fischer.
Wiener, A.S. 1970. Immunochemical studies on groupspecific agglutinins of diverse origins. *Haematologia* 4: 157–166.
Wilson, A.C., White, T.J., and Carlson, S.S. 1977. Molecular evolution and cytogenetic evolution, in: *ICN-UCLA Symp. Molec. and Cell Biol.* VII: 375–393. eds. R.S. Sparks, D.E. Comings, C.F. Fox.
Wilson, D.R. 1972. Tail reduction in *Macaca*, in *The functional and evolutionary biology of primates*. ed. R.H. Tuttle, 241. Chicago, New York: Aldine Atherton.
Winge, H. 1895. *Jordfundne og nulevende Aber (Primates)*. Kopenhagen: E Musei Lundi.
Wolin, L.R., and Massopust, G. 1970. Morphology of the primate retina, in *Advances in primatology I*. eds. C.R. Noback, and W. Montagna, 1. New York: Appleton Century Crofts.
Wood-Jones, F. 1929. *Man's place among mammals*. Arnold, London.
Wrobel, K.H. 1966. Untersuchungen an den Blutgefässen des Primaten Greifschwanzes. *Z. Anat. Entwicklungsgesch.* 125: 177.

Yunis, E.J., deCaballera, O.M.T., Ramerez, C., and Ramerez, Z. 1976. Chromosomal variation in the primate *Alouatta seniculus seniculus*. *Folia Primatol.* 25: 215–224.

Yunis, E., deCaballero, O.M.T., and Ramerez, C. 1977. Genus *Aotus* Q- and G-band karyotypes and natural hybrids. *Folia Primatol.* 27: 165–177.

Yunis, J.J., Sawyer, J.R., and Ball, D.W. 1978. The characterization of high-resolution G-banded chromosomes of man. *Chromosoma* 67: 293–307.

Zuckerman, S. 1932. *The social life of monkeys and apes.* London: K. Paul Trench Trubner.

Index

Page numbers in boldface italic refer to illustrations.

A

Acceleration, 6
Airorrhynchy, 128
Airsac, 68, 69
Allen's swamp monkey. See *Allenopithecus*
Allenopithecus, 61
Allocebus, 16
Allometry, 130, 137, 147
Alouatta, 38, 41, 42, 43, 105, 116, 127, ***128***, 162, 175, ***181***, 181, 199, 203, 250, 256, 271
 dentition, 80
 palliata, ***42***
Alouattinae, 38, 41, 148, 225
Altner, G., 189, 193, 242
Alveoli, 77, 85
Analogy, 5, 8
Anaptomorphidae, 99
Ancestors, 5, 7, 8
Angwantibo. See *Arctocebus*
Ankel, F., 148, 164, 199
Ansa coli, 226
Anthropoidea, 15, 36, 37, 76, 100, 114, 121, 127, 130, 147, 148, 162, 202, 205, 206, 210, 212, 221–23, 233, 237
Anthropoideans, 10, 114
Anthropomorpha, 3

Aotinae, 38
Aotus, 38, 39, 85, 114, 115, 147, 148, 172, 181, 219, 221, 223, 271
Apes, 1–4, 6–8, 10, 15, 36, 45, 67, 69, 70, 75, 76, 154, 178, 238
 African, 76
 greater, 237
Arboreal, 11, 16, 26, 33, 42, 46, 51, 53, 63, 136, 245
 quadrupeds, 252, 253
Arctocebus, 30–32, 76, 161, 162, 172, ***196***
 dentition, 96
 skull, 126
Arctopithecus, 47
Arm swinging, 5, 45, 68, 183, 245, 250; *see also* Brachiation
Arteria vertebralis, 153
Ashton, E. H., 246–48
Ateles, 38, 44, 45, ***105***, 148, 149, 155, ***162***, 175, ***181***, 181, 190, 199, 235, 236, 248, 255, 256, 271
 thumb reduction, 191
Atelinae, 38, 44, 148, 183
Australopithecus, 6
Avahi, 15, 24, 25, 158, 185, 220, 248, 252
 dentition, 95
Aye aye, 16, 28; *see also Daubentonia*

303

B

Baboons. *See Papio*
Baculum, 235, 236
Badrian, A., 71
Badrian, N. L., 71
Barbary ape. *See Macaca sylvanus*
Bauchop, T., 146, 225
Bearded saki. *See Chiropotes*
Behavior, 7, 11, 43, 46, 53, 59, 71, 75
  baboon, 56
  comparative, 6
  human, 74
  macaque, 53
  marking, 19, 21, 22
  primate, 74
  variation, 5
Bender, M. A., 258
Bennejeant, C., 79
Biegert, J., 129, 130, 188
Bierens de Haan, J. A., 220
Bilophodont, 101, 104, 106, 107, **108, 109,** 109
Biology, 1, 4
  of modern humans, 75
  of primates, 5
Bipedalism, 46, 53, 56, 62, 183
  bipedal hopping, 34
  bipedal running, 53
  bipedal walking, 43, 68, 75, 254, 256
Birth canal, 177, 178
Birth, 177, 242
  season, 26, 43, 57, 237, 238
  single, 38, 43, 47, 240
  see also Littersize
Black mangabey. *See Cercocebus aterrimus*
Blaffer-Hrdy, S., 65
Bleeding heart baboon. *See Theropithecus*
Bluntschli, H., 208, 209
Body size, 9, 12, 17, 20, 21
Bolk, L., 6, 78
Bonobo. *See Pan paniscus*
Boswell, J., 75
Brachiation, 245, 250–52
Brachiators, 5, 253
Brachyteles, 38, 44, 45, 106, 181, 190, 199
  thumb reduction, 191
Breeding, 40, 41, 62
  season, 20, 21, 24, 34, 36, 39, 41, 43, 45, 46, 65, 68, 72, 74, 237
Brues, A., 75
Buettner-Janusch, J., 274
Bulla, 119, 120, 122, 126
Bushbabies. *See Galago*

C

Cacajao, 38, 40, 41, 104, 105, 164, 172, 181
Callicebinae, 38
Callicebus, 38, 39, 175, 207, 208, 263, 271
Callimico, 46, 47, 104, 267, 271
Callimiconidae, 38
Callithrix, 47, 48, 86, 175, **181, 195,** 267, 276
  brain, **144**
  dentition, 80, 81, **106,** 106, 107
  jacchus, **50,** 189, 195
Callitrichidae, 38, 47–49, 51, 81, 100, 106, 107, 129, 147, 148, 164, 181, 184, 192, 208, 227, 235, 267, 271, 275, 276
Capuchin monkey. *See Cebus*
Carnivora, 9, 169
Carotid artery, 122
Carpenter, C. R., 43, 68
Cartilage replacement bone, 113
Cartmill, M., 16, 115, 120, 123, 176, 249
Casperson, T., 261
Castenholz, A., 219
Catarrhini, 37, 117, 129, 131, 178, **205,** 205, 211
Cebidae, 38, 43, 81, 129, 148, 159, 162, 193, 203, 208, 227, 229, 237, 238, 271
  dentition, 104, **105,** 105
Ceboidea, 36–38, 44, 162, 195, 267
Cebuella, 23, **49,** 49, 87, 106, 148, 158, 210, 236, 267
Cebus, 38, 44, 106, 121, 135, 172, 181, 199, **213,** 256, 271
Celebes black ape. *See Cynopithecus*
Cellulose, 88
Cercocebus, 59, 76, 223, 242, 272
  aterrimus, 60
Cercopithecidae, 51, 182, 183, 203, 213, 227, 237
Cercopithecinae, 51, 62, 63, **131,** 131, 164, 225, 240
Cercopithecoidea, 36, 37, 51, 62, 129, 162, 165, 184, 208, 210, 229, 272
  carotid artery, 123
  dentition, **101,** 100–104, 107, 109
Cercopithecus, 47, 59–61, 63, 76, 101, 187, 236, 238, 242, 272, 275
  aethiops, **61,** 238
  dentition, **101,** 104
Chalmers, N. R., 60
Charles-Dominique, P., 32, 33, 36, 87
Cheek pouches, 51, 52, 62, 225
Cheirogaleidae, 15, 16

Index

Cheirogaleinae, 16, 195
  carotid artery, 123
  dentition, 95
Cheirogaleus, 15, 23, 202, 220, 226, 236, 241
  dentition, 96
  major, 22
  medius, 22
Chevron bones, 157
Chimerism, 275, 276
Chimpanzee. *See* Pan
Chiropotes, 38, 40, 104, 105, 181, 256
Chiroptera, 9, 100
Chitin, 88
Chivers, D. J., 68
Christen, A., 50
Chu, E. H. Y., 258
Cingulum, 85, 88, 90, 93, 94, 100, 110
Circle of Willis, 123
Clark, W. E. Le Gros, 4
Classification, 37
Claws, 10, 14, 28, 36, 47, 48, 187, **188**, 188, 192
  toilet, 16, 36, 195
Climber, 5, 43, 45, 46, 53, 70, 73, 169, 188, 213
Clinging. *See* Vertical clinging and leaping
Clitoris, 44, 45, 236, 237
Coimbra-Filho, A. F., 87
Collias, N., 203
Colobidae, 227
Colobinae, 42, 51, 62, 63, 66, 76, 102, 130, 131, 148, 182, 204, 225, 252, 272
  thumb reduction, 191
Colobus, 62, 63, 76, 135, 155, 204, 236, 238, 239, 255, 272
Conroy, G. C., 120
Cope, E. D., 86
Copulation, 24, 39, 72, 240
Crepuscular, 12, 15, 219
Cretaceous, 9, 10, 203
Crick, F. H. C., 276
Crista obliqua, 108, 109
Cuvier, G., 2, 126
Cynomorpha, 55, 116
Cynopithecus, 53, **54**, **216**, 272

D

Dahlberg, A. A., 77
Darwin, 1, 3, 215, 277
Darwinian angle, 52, 215, **216**
Daubentonia, 15, **28**, 28, 95–98, **97**, 116, 146, 147, 192, **193**, 208, 217
  skull, 126

Daubentoniidae, 15, 96, 117, 158
  skull, 125
Deciduous teeth, 78; *see also* Milk dentition
Dendrogale, 12
Dental arcade, 112
  snout, 130, 203
Dermal bone, 113–16, 169
Dermatoglyphics, 16, **192**, 253
Dexterity, 13, 29
Diastema, 81, 82, 89, 90, 92, 93, 104, 111
Dichromatism, 18, 42, 67
Digitigrade. *See* Walking
Dimorphism, 57
  color, 42, 63, 67
  sexual, 18, 52, 55, 63, 66, 68, 69, 71, 73, 84, 102, 109, 111, 178
Distal
  defined, 12
  in teeth, 78
Diurnal, 12, 20, 25, 26, 31, 39, 41, 44, 46, 48, 54, 68, 71, 74, 129
DNA, 260, 262, 263, 274, 276
Dolichopithecus, 129
Doran, A. H. G., 217
Douc langurs. *See* Pygathrix
Douroucouli. *See* Aotus
Doyle, C. A., 36
Drill. *See* Mandrillus
Dryopithecus, 107
  pattern, 107, **108**, 108, **109**, 109, 111
Dukelow, W. R., 242
Dutrillaux, B., 274
Dwarf lemurs. *See* Cheirogaleus

E

Ear, **214**, **215**
  crossfolds, **34**, 211
  inner, 10
  membranous, 22, 24, 29, 30, 33, 35, 211, 213
  middle, 10
  mobile, 12, 24, 35
  ossicles, 113
Eastern gorilla, *Gorilla gorilla beringei*, 73
Eastern lowland gorilla, *Gorilla g. manyema*, 73
Edinger, T., 137
Electromyography, 256, 257
Elephant shrew (*Macroscelides*), 1
Ellefson, J. O., 68
Entepicondylar foramen, 172, **173**, 175
Eocene, 88

Epple, G., 202, 236
Erikson, G. E., 43, 163
Erskine, A. G., 280
Erythrocebus, 62, 76, 101, 104, 242, 253
Estrus, 52, 55, 61, 72, 238, 239
Euoticus, 22, 23, 33
  dentition, 96
Evans, E. P., 276
Every, G., 102
Evolution, 3
  evolutionary achievement, 16
  human, 4, 6, 75
  language, 75
  organic, 7
  primate, 75
Extinction, 8, 9, 51, 74
Eye
  color, 18
  frontality, 12, 25
  laterally directed, 14
  lids, 223, 224
  sockets, 11, 15

F

Fiedler, W., 37
Fissure of Rolando, 143
Fleagle, J. G., 49, 68, 241, 248, 255, 256
Fooden, J., 52
Ford, C. E., 259, 276
Forehead, 11, 41, 44, 149
Fossa canina, 134
Fossey, D., 74
Fossils, 5, 6, 9, 77, 81, 107, 122, 149, 277
Friant, M., 94
Friction pad, 35, 42, 44, 46
  skin, 193, 199
Frima, M. J., 220
Frontal, 11

G

Galagidae, 15, 76, 178, 213, 220, 252
Galaginae, 158
Galago, 16, 33, 146, 194, 200, 201, 210, 213, 214, 248, 249
  crassicaudatus, **34, 196,** 282
  demidovii, 23, 50, 233
  dentition, 95, 96
  elegantulus, 87
Galagoides, 33
Galdikas-Brindamour, B., 70
Gartlan, J. S., 62
Garver, J. J., 274
Gasden, J. R., 274

Gelada. See *Theropithecus*
Gentle lemur, 21; see also *Hapalemur*
Geoffroy, E., 126, 200
Gérard, P., 233
Gestation length, 241, 242
Gibbons. See *Hylobates*
Glander, K., 42, 43
Glands, 202, 236
  axial, 22
  circumanal, 36, 38
  cutaneous, 19, 26
  forearm, 21
  salivary, 208
  throat, 40
Glaser, D., 137, 203, 210, 211, 240
Goeldi's marmoset. See *Callimico*
Goodall (van Lawick), J., 71, 72, 74, 75, 225, 251
Gorilla, 5, 72, 74, 84, 134, 170, **175,** 182, **184,** 184, 185, 187, **192, 194,** 235, 240, 250, 273, 283
  brain, **144**
  g. beringei, **73**
  g. gorilla, **73**
  mastoid, 120
  skull, **132, 133**
Gould, S. J., 137
Grand, T. I., 241
Greater gibbon. See *Symphalangus*
Gregory, W. K., 77, 88, 97, 107, 245
Grooming, 16, 30, 36, 39, 52, 57, 60, 68, 72, 74, 87
Grosser, O., 231
Group. See Social group
Groves, C. P., 73
Guareza. See *Colobus*
Guenons. See *Cercopithecus*
Gustatory effects, 210, 211

H

Haeckel, E., 37
Haines, R. W., 114
Halaczek, B., 180, 181, 182
Hallux, 16, 47
  abducted, 16
Hamerton, J. L., 258
Hamilton, A. E., 274
Hapale, 47
Hapalemur, 16, 20, 21, 202, 220, 248
  griseus, 21
  simus, 21
Hapalidae, 47; see also Callitrichidae
Hapanella, 47
Haplorhini, 202, 205, 206
Harem, 57
Harlow, H. F., 239

Index

Harlow, M. K., 239
Hearing, 218
Heffner, H. E., 218
Hellekant, G., 210
Hellman, M., 88
Hemprich, W., 205
Hennig, W., 283, 285
Hershkovits, P., 37, 47, 48
Herwerden, M. van, 233
Hesse, R., 212
Heterodont, 77
Higher primates. See primates
Highland or Eastern gorilla. See *Gorilla g. beringei*
Hiiemae, K. M., 86, 87, 88
Hill, W. C. O., 10, 51, 77
Histology, 160
  of brain, 137, 145, 147
  of teeth, 78, 97
  of small intestine, 225
Hofer, H. O., 117, 135, 137, 148, 204, 207, 208, 242
Holloway, R. L., 137
Hominidae, 74, 182
  tooth pattern, 112
Hominids, 178
  earliest, 5
Hominoidea, 36, 37, 67, 68, 130, 143, 148, 154, 162–65, 167, 169, 206, 210, 227, 229, 232, 240, 245
  carotid artery, 123
  tooth pattern, 107, 109, 111
Homo, 2–4, 7, 8, 74–76, 117, 129, 136, 148, 149, 154, 175, 179, 182–85, **184, 194, 198,** 198, 212, 218, 222, 229, 237, 240, 244, 254, 278, 284, 285
Homo erectus, 149
Homo habilis, 149
Homology, 147, 208
  of chromosomes, 262
Hone, 87, 102, **103,** 105
  in Hylobatidae, 109
  in Pongidae, 111
Horn, D. A., 71, 251
Howler monkey. See *Alouatta*
Human evolution. See Evolution
Humans, 1–7, 10, 15, 36, 48, 51, 53, 74; see also *Homo*
Humeral torsion, 172
Huxley, T. H., 1, 3
Hybridization, 275
Hylander, W. L., 86
Hylobates, 5, 67, 68, 76, 85, 116, 134, 172, 183, 187, **194,** 223, 240, 256, 272, 273
  lar, **67**

Hylobatidae, 182, 184, 187, 235
  dentition, 109, 110
Hyoid, 39, 42, 113, 120
  enlarged, 127, **128**

I

Incisura ischiadica, **174,** 178
Indri, 25, 26, 162, 185, 187, 220, 226, 248, 252
  dentition, 95
Indriidae, 15, 26, 116, 121, 123, 147, 158, 195, 252
  dentition, 80, 81, **94,** 95
  skull, 125
Infanticide, 65
Insectivora, 9–11, 100, 138, 144, 145, 209, 229
Intermembrane bone, 113
Iodopsin, 218
Ischial callosities, 52, 55, 56, 58–61, 63, 68, **175,** 178, 236

J

Jacchus, 47
James, M. M., 77, 80
Jenkins, F. A. Jr., 161, 170, 257
Jolly, A., 26, 74, 225, 229
Jolly, C. J., 224, 238, 253
Jones, M. L., 241

K

Karyotype, 259–61, 266, 272, 273, 275, 277
Kay, R. F., 86–88
Keith, A., 245
King, M. C., 274
Knuckle walking. See Walking
Kuhn, H. J., 62, 225, 236, 239

L

Lagothrix, 38, 44–46, **46,** 162, **171,** 181, 199, **205,** 235, 271
Lamina papyracea, 121
Landsteiner, K., 279
Langurs. See *Presbytis*
Laryngeal apparatus, 42
  sac, 27
Lasinsky, W., 212, 215
Latistarnalia, 155
Lawick, J. van. See Goodall
Leaf monkeys. See *Presbytis*
Leaping. See Vertical leaping
Leche, W., 80

Lemur, 15, 16, 91, **92,** 95, 117, **122,** 126, **144,** 145, **190,** 194, 202, 213, 267, 274, 282
  catta, 17, **18,** 20, 220, 237, 240, 249, 252
  fulvus, **18**
  fulvus albifrons, **20**
  macaco, 18
  mongoz, 211
  skull base, **124, 125**
  variegatus, **17;** see also *Varecia*
Lemuridae, 15, 16, 99, 120, 121, 123, 147, 158, 195, 203, 208, 220, 223, 226, 229, 232
  skull, 125
Lemuriformes, 37, 91, **131,** 207, 217
Lemurinae, 16
Lemuroidea, 16
Lemurs, 3, 8, 11, 75, 130
  mongoose, 19
  redbellied, 19
  ruffed, 19, 20
  subfossil, 16
Leontideus, 48, 267
Leontocebus, 47, 172, 195
Leontopithecus, 48, 50, 51
Lepilemur, 15, 16, 21, 83, **84,** 84, 91, 117, 121, 185, 248, 263, 266, 271
Lepilemuridae, 15
Lesser apes. *See Hylobates*
Lesser gibbons. *See Hylobates*
Lesser mouse lemur. *See Microcebus*
Leutenegger, W., 148, 163, 177, 241, 242
Levan, A., 258
Liebermann, P. H., 75
Liocephalus, 47
Lion tamarin. *See Leontopithecus*
Lissencephaly, 143, 145
Littersize, 49; see also Twin birth
  ruffed lemur, 21
  treeshrews, 14
Locomotion, 5, 31, 44, 46, 49, 60, 68, 75, 148, 165, 169, 172, 177, 178, 180, 183, 241, 242
  armswinging, 176
  classification, 26, 249
  leaping, 49, 50
  specialized, 36, 195, 196
Locomotor behavior. *See* Locomotion
Lorenz, R., 189, 195, 202
Loris, **29,** 29, 30, 146, 161, 220
  pelvic symphysis, 177
  skull, 126
  tympanic, 120
Lorisidae, 15, 29, 30, 120, 121, 126, 158, 159, 189, 195, 203, 206, 213, 220, 226, 229, 232, 237

carotid artery, 123
dentition, 95, 96
retia mirabilia, 191
Lorisiformes, 37, 91, 207, 237
Lorisinae, 160, 161
Lovejoy, C. D., 277
Lowland or western gorilla. *See Gorilla g. gorilla*
Luckett, W. P., 10, 11, 233
Lyon, M. W., 15

M

Macaca, 51, 52, 60, 76, 104, 121, **122,** 130, **162,** 172, **173, 175, 194,** 214, 223, 224, 236, 238, 239, 242, 272, 275, 280, 282
  brain, **144**
  mastoid, 120
  skull base, **124**
Macaca silenus, 53
  sylvana, 51, 76
Macaques. *See Macaca*
Maclatchy, A., 60
MacPhee, R., 120
Malbrant, R., 60
Mammals, 1, 3, 9–11, 72, 75, 136, 138, 168, 169, 206, 210, 212, 220, 224, 244
  extant, 86
  nonprimate, 15, 16
  placental, 9, 228
  primitive, 15, 123, 207
  teeth of generalized, 78, 223
Man. *See* Humans
Mandibular symphysis, 11, 127
  arcade, 125
Mandrill. *See Mandrillus*
Mandrillus, 54, 235
Mangabeys. *See Cercocebus*
Marikina, 48
Marking, 20, 21, 24, 36, 202
  scent, 19, 21, 203, 238
  throat, 26
  urine, 19, 26, 30, 33, 36, 38, 202
  see also Behavior
Marmosets. *See* Callitrichidae
Martin, R. D., 36, 235
Martucci, R. W., 225
Masali, M., 120
Massopust, G., 219
Mastication, 87, 88
Mastoid, 120
Mating season, 20, 24, 26
McArdle, J. E., 246, 249
Membrane bone. *See* Intermembrane bone
Mendel, F., 248, 250, 257

Mesial, 78
Mesostyle, 106
Metatarsal elongation, 197
Metopic suture, 11, 119
Mico, 47
Microcebus, 15, 16, 23, 24, 121, 138, 145, 146, 194, **196,** 200, **201,** 201, 202, 213, 214, 220, 223, 226, 236, 238, 240
    dentition, 96
    Mirza coquereli, 23
    murinus, **23,** 23, 24, 50
    rufus, 24
Midas, 48
Military monkey. *See Erythrocebus*
Milk dentition, 78–81
    Daubentonia, 97
    Indriidae, 94
    Lepilemur, 91
Milton, K., 176
Miocoella, 47
Miopithecus, 39, 61, 104, 239
Mirza, 23, 24
    coquereli, 23
    dentition, 96
Mittermeier, R. A., 49, 241, 248, 255, 256
Mivart, S. G., 37
Mob call, 21
Molecular biology, 6
    clock, 278, 279
    research, 4
Monellin, 211
Monk saki. *See Pithecia*
Monkeys, 1, 2, 3, 5, 7, 8, 15, 36, 53, 59, 65, 154
    baboon like, 57
    newborn, 7
    smallest, 49
Monophyletic, 37
Morbeck, M. E., 246, 248, 257
Moskowitz, N., 223
Mouse lemurs. *See Microcebus; Mirza*
Multiple birth, 13
Mutation, 262
Mystax, 48

N

Nails, 9, 10, 16, 48, 187, **188,** 188
    keeled, 22, 23
    modified, 48
    true, 196
Napier, J. R., 188, 189, 203, 246–48, 251, 252
Napier, P. H., 188, 203, 246–48, 251
Nasal opening, 71, 73
    pipelike opening, 126

septum, 37, 41–45, 48, 52, 53, 60, 117
    wings, 36, 37, 117
Nasalis, 66, 67, 182, 187, 204, 240, 242, 272
Needle clawed bush baby. *See Euoticus*
Nest, 12, 14, 21, 24, 29, 34, 70, 72, 74
    termite, 75
Night monkey. *See Aotus*
Nishida, T., 72
Noback, C. R., 223
Nocturnal, 12, 15, 21, 24, 30, 38, 60, 147, 213, 218, 221
Nose, 203–205
    fleshy, 66
    snub, 55, 66
Notharctus, 88
Nycticebus, 30, **31,** 75, 146, 161, 220, 242, 267, 274
    tympanic, 120

O

Odontoceti, 138
Oedipomidas, 48, 236
Oedipus, 48
Ohno, S., 263
Olfactory messages, 19
    muzzle, 203
Oligocene ceboid, 129
Ontogeny, 6, 78, 113, 119, 123, 214, 228, 229, 230, 231, 233, 237
    of auditory region, 120
    of brain, 139, 141
    of feet, **185,** 185, **198,** 199
    of hands, **185,** 185
Orangutan. *See Pongo*
Orbit, 9
    frontality, 10, 16
    postorbital closure, 11
Orientation of teeth, **78,** 78
Os incisivum, 82
Osborn, H. F., 86
Ouistitis, 48
Owen, R., 2, 3
Owl monkey. *See Aotus*
Oxnard, C. E., 246–48, 257

P

Pair bond, 19, 20, 24, 38, 39, 47–49, 51, 53, 68
Pan, 2, 5, 68, 70, 74, 75, 111, 134, **171,** 182, 183, 185, 216, 218, 237, 239, 242, 256, 273, 283
    mastoid, 120
    paniscus, 70, 71, **72,** 251
    skull, **132, 133**

Papio, 53–57, 60, 66, 76, 84, 102, 104, **114**, 121, 130, 177, 197, 214, 224, 236, 238, 242, 272, 275
  cynocephalus, **56**
  mastoid, 120
Parallelism, 148
Patas monkey. *See Erythrocebus*
Pearson, P. L., 274
Pelycodus, 88
Penis, 10, 44, 235, 237
Pentailed tree shrew. *See Ptilocercus*
Pericone, 93, 94
Perodicticus, **32**, 32, 33, 76, 155, **161**, 165, 172, **190, 201**, 201, 202, 208, **214**, 235, 236
  cervical spines, 159, **160**
  dentition, 96
  skull base, **124**
Peters, W., 29
Petter, J. J., 26, 266
Petter-Rousseaux, A., 238
Phaner, 15, 16, 22, 23, 201, 213
  dentition, **83**
  furcifer, 22, 87
Phanerinae, 16
Phenylthiocarbamid. *See* PTC
Pheromones, 238
Philippine tree shrew. *See Urogale*
Philtrum, 200
Pilbeam, D. R. 74
Piliocolobus, 62; see also *Colobus*
Pithecia, 38, 40, 41, 104, 105, 172, **173**, 256
Pithecinae, 38, 40, 148
Platyrrhini, 37–39, 46, 48, 51, 117, 120, 129, 131, **205**, 205
Pocock, R. I., 202
Pollock, J., 27, 28
Pongidae, 129, 130, 149, 178, 182, 183, 203, 227
  dentition, 110–12
Pongo, 68, **69**, 69, 70, 73, 76, 85, 134, 182–84, **194**, 205, 256, 273, 285
  dentition, 110, 111
  forehead, 149
  mastoid, 120
  skull, **132, 133**
Postorbital constriction, 126
Potto. *See Perodicticus*
Prehensile tail, 42, 44–46, 136, 148, 157, 163, 164, 199, 252
Presbytis, 63, 65, 66, 76, 238, 255, 272
  entellus, **65**
  obscurus, **64**
  thumb reduction, 191
Primate distribution, **13**

Primates, 1–3, 9, 10, 50
  extant lower, 16
  higher, 10, 15, 16
  living, 5
  non-hominid, 75
  smallest, 50
  typical lower, 16
Primatology, 1, 2, 4, 8
Proboscis monkey. *See Nasalis*
Procolobus, 62; see also *Colobus*
Procumbent teeth, 40, 50, 82, **83**, 83, 84, 90, 95, 102, 105, 110, 208; see also Tooth comb
Pronation, 175
Propithecus, 25, 83, 146, 158, 187, 220, 224, **226**, 226, 248, 252
  dentition, **94**, 95
  diadema, 25
  verreauxi, **25**, 25, 27
Prosimians, 1, 10, 15, 16, 36–38, 130, 145–48, 154, 162, 163, 165, 177, 204, 209, 210, 212, 227
  largest living, 26
  tooth comb, 95, 99
Protocristid, 106, 107
Protostyle, 94
Proximal, defined, 12
Pseudohypocone, 88
PTC, 210
Ptilocercinae, 11
Ptilocercus, 11–12, 75, 145, 212, 213
  lowii, 11, 12
Purgatorius, 9
Pygathrix, 65, 76, 272
Pygmy chimpanzee. *See Pan paniscus*
Pygmy marmoset. *See Cebuella*

Q

Quadrupedal, 18, 22, 34, 41, 52, 58, 60, 62, 179
  climbing, 42, 44, 73
  leaper, 38, 39, 43, 50
  runner, 5, 39, 43, 49
  walking, 49, 56, 70

R

Radinsky, L., 137, 143
Reduction, pollex, 190, 191
  premaxilla, 84, **84**, 117
  tooth, 81
Remane, A., 77, 80, 81, 82, 83, 94, 98, 108, 111, 277
Rempe, U., 285
Retardation, 6
Rete mirabile, 191
Reynolds, F., 71, 72

Index

Reynolds, V., 71, 72
Rhinarium, 13, 16, 35, 200, **201**, 201, 202
  defined, 13
Rhinopithecus, 66, 76, 204
Rhodopsin, 218
Richard, A., 26
Riopelle, A. J., 242
Rodent like, 121, 126
Rodentia, 9, 28, 96, 98
Rodman, P. S., 70
Rohen, J. W., 219
Rose, M. D., 251, 257
Ruch, T. C., 4
Ruffed lemur. *See Varecia*
Rumpler, Y., 15, 16, 274

S

Saban, R., 123
Sacculated stomach, 225
Sacred baboon. *See* Papio
Sagouin, 48
Saguinus, 47, 48, 107, 210, 256, 267
Saimiri, 38, 43, 44, 128, 129, **130**, 172, 178, **181**, 195, 210, 226, 238, 256, 271
Saimirinae, 38
Saki. *See Pithecia*
Sarich, V. M., 277
Scaled, 12
Scent glands. *See* Glands
Schaller, G. B., 74, 204, 251
Schneider, R., 208, 209
Schultz, A. H., 4, 111, 161, 177, 215, 223, 235, 239, 241, 242
Schwartz, J., 94, 98, 99
Sectorial, 103, 105, 111
Seniocebus, 48
Sense organ, 16
Sensory skin, 13
Sexual dimorphism. *See* Dimorphism
Siamang. *See Symphalangus*
Sifaka. *See Propithecus*
Silver backed males, 74
Simia, 48
Simiae, 37
Simias, 48, 66, 76, 204
Simii, 36
Simons, E. L., 37, 74, 75
Simopithecus, 252
Simpson, G. G., 37, 283
Skull base, 116, **125**
  crest, 119, 135
Slender loris. *See Loris*
Smooth tailed tree shrew. *See Dendrogale*
Snub nosed langur. *See Rhinopithecus*

Socha, W. W., 280
Social, 21, 236
  behavior, 43, 46, 56
  contact, 16
  groups, 19, 20, 23, 24, 41, 43–49, 53, 56, 57, 60, 62, 63, 65, 66, 71, 72, 74
  interactions, 40
  signal, 18
South American monkeys, 36, 49
Southwick, C., 203
Southwick, G. H., 53
Spatz, W. B., 120
Spider monkey. *See Ateles*
Sportive lemur, 21; *see also Lepilemur*
Sprankel, H., 126, 146, 199
Spreng, H., 80
Squirrel monkey. *See Saimiri*
Stages of Life, 240
Stapedial artery, 11, 123
Starck, D., 137, 143, 146, 148, 149, 227, 228, 232, 233, 235
Staring, 18, 19, 26
Stehlin, H. G., 88, 94
Steiner, H., 186
Stephan, H., 137, 146
Stern, J. T., 256, 257
Stomach, 51
  sacculated, 43, 62
Strepsirhini, 200, 202, 205, 206
Sublingua, 207–209, **209**
Sublingual organ, 207
Suction cups, 191
Sugiyama, Y., 71
Supination, 175
Susman, R. L., 256, 257
Suzuki, A., 71
Swamp monkey. *See Miopithecus*
Swimming, 60, 70
Swindler, D. R., 77, 94, 106–108, 199
Sylvanus, 48
Sylvian fissure, 143, 145, 147, 148
Symphalangus, 68, 76, **134, 160, 171,** 187, 273
Symphysis, 11, 115
  of pelvis, 177
  see also Mandibular symphysis
Syndactyly, 24, 26, 31, 66, 68, 187

T

Tactile pads, 136, 190
  nerves, 160
  ridges, 201
  sense, 147, 200
  skin, 16, 164

Tail. *See* Prehensile tail
Talapoin monkey. *See Miopithecus*
Talonid, 86, 90, 93, 96, 99, 100, 103, 111
Tamarin, 48
Tamarins. *See* Callitrichidae; *Saguinus*
Tamarinus, 48
Tapetum lucidum, 220, 221
Tarsal elongation, **196**, 197, 249
Tarsiers, 8, 36, 76, 189, 191, 252
Tarsiidae, 15, 34, 35, 158, 178, 201, 220, 229, 232, 235
  carotid artery, 123
Tarsiiformes, 37, 217, 226
Tarsius, 16, 34, **35**, 35, 76, 82, 86, 87, 114, **115**, 115, 116, 121, 123, 129, 137, 146, 148, 155, 158, **159**, 159, 165, 169, 189, **190**, 194, **196**, 197, 199, 206, 208, 210, 214, 220, 221, 224, 232, 233, 237, 248, 249, 263, 275
  bancanus, **35**, 126, **127**, 127
  dentition, **98**, 98, 99
  eye volume, 126
  tympanic, 120
Tattersall, I., 23, 98
Taxonomy, 3, 6, 37, 47, 117, 120, 137, 205, 217, 228, 232, 276, 283
T-beam principle, 170
Terrestrial, 16, 53, 56, 57, 59, 73, 176, 178, 252, 253
  large mammals, 203
Territorial calls, 21, 25, 42, 43
Territory, 20, 26, 30, 38, 39, 40, 41, 44, 46, 53, 60, 63, 68, 70
Thaumatina, 211
Thenius, E., 285
Theropithecus, 57, **58**, 58, **59**, 76, 130, **160, 171**, 224, 252, 254, 272
Threat, 39, 69, 87, 224, 236
Thumb
  abductable, 19
  opposable, 52
  reduced, 45
Tigges, J., 145, 219
Toe, big, 10, 16, 31, 48
Toilet claw. *See* Claws
Toolmaker, 75
Tooth comb, 82, **83**, 83, 87, 89, 91, 92, 95, 105
  prosimians, 95, 99
Torpor, 22, 24
Tree shrews. *See Tupaia*
Trigonid, 86, 88, 91, 93, 96, 99, 100, 104
Trituberculum, 86

Tuberculum Darwini, 215
Tupaia, 1, 10, 12, **14**, 14, 15, 75, **89**, 89, 117, 130, **144**, 145, 189, **190**, 193, 194, 197, 212, **213**, 213, 219, 220, 232, 233, 266
Tupaiidae, 227, 235
Tupaiiformes, 10, 11, 37, 120, 123, 124, **131**, 188, 217, 226, 229, 237
Tupainae, 11, 12, 223
Tupaioidea, 10, 11
Tuttle, R. H., 71, 191, 250
Twins, 45, 48–51, 240, 275; *see also* Littersize
Tyson, E., 2

U

Uakari. *See Cacajao*
Upright, 3, 68, 127, 165, 178
  jumping, 249
  sitting, 20
  walker, 5, 71, 129, 251, 254
Urine washing. *See* Marking
Urogale, 12, 219
Usurper strategy, 65

V

Van Valen, L., 10
Varecia, 16, **17**, 20, 202, 210, 236
  variegata, 20
Variability, 5, 6, 9, 107, 111
Vertebra prominens, 165
Vertical clinging and leaping, 24, 25, **35**, 35, 36, 47, 127, 129, 179, 195, 248, 249, 252
Vertical position, 21, 26, 33, 159
Vervets. *See Cercopithecus*
Visual axes, 15
Vision, 113
  binocular, 15
  field of, 15, 16
  stereoscopic, 16
Vocalization, 25, 26, 36, 38, 39, 42, 44, 49, 50, 51, 60, 68, 69, 71, 72, 74, 206
Vogel, C., 135
Vogt, G., 148
Vogt, O., 148

W

Walker, A., 30, 33, 160, 252
Walking, 199; *see also* Bipedal walking
  digitigrade, 56, 60, 62, 191
  fist, 69

Index

knuckle, 71, 74, 183, 191, 250, 251, 254
plantigrade, 70
see also Upright walking
Wanderoo. *See Macaca silenus*
Watson, J. O., 276
Webbing. *See* Syndactyly
Wen, I. C., 205
Werner, C. F., 123, 217
Western gorilla. *See Gorilla g. gorilla*, 73
Wettstein, E. B., 107
Wible, J., 120
Wiedersheim, R., 207
Wiener, A. S., 279, 280
Wilson, A. C., 274, 278, 279
Wilson, D. R., 52

Winge, H., 97
Wolin, L. R., 219
Wood, C. D., 199
Wood-Jones, F., 245
Woolly indri. *See Avahi*
Woolly monkey. *See Lagothrix*
Woolly spider monkey. *See Brachyteles*
Wrobel, K. H., 148

Y

Yunis, E. J., 262

Z

Zuckerman, Sir S., 237